Synthesis Lectures on Data Mining and Knowledge Discovery

Series Editors

Jiawei Han, at Urbana-Champaign, University of Illinois, URBANA, IL, USA

Lise Getoor, University of California, Santa Cruz, Santa Cruz, USA

Johannes Gehrke, Microsoft Corporation, Redmond, WA, USA

The series focuses on topics pertaining to data mining, web mining, text mining, and knowledge discovery, including tutorials and case studies. Potential topics include: data mining algorithms, innovative data mining applications, data mining systems, mining text, web and semi-structured data, high performance and parallel/distributed data mining, data mining standards, data mining and knowledge discovery framework and process, data mining foundations, mining data streams and sensor data, mining multi-media data, mining social networks and graph data, mining spatial and temporal data, pre-processing and post-processing in data mining, robust and scalable statistical methods, security, privacy, and adversarial data mining, visual data mining, visual analytics, and data visualization.

Jiaming Shen · Jiawei Han

Automated Taxonomy Discovery and Exploration

 Springer

Jiaming Shen
Google Research
New York City, NY, USA

Jiawei Han
University of Illinois Urbana-Champaign
Urbana, IL, USA

ISSN 2151-0067   ISSN 2151-0075   (electronic)
Synthesis Lectures on Data Mining and Knowledge Discovery
ISBN 978-3-031-11407-6   ISBN 978-3-031-11405-2   (eBook)
https://doi.org/10.1007/978-3-031-11405-2

This Springer imprint is published by the registered company Springer Nature Switzerland AG
The registered company address is: Gewerbestrasse 11, 6330 Cham, Switzerland

To my family for their love and support.

Preface

In today's information era, people are inundated with vast amounts of text data. Every day, there are thousands of scientific papers, tens of thousands of news articles, corporate reports, and millions of social media posts produced and shared worldwide. Turning those massive text data into actionable knowledge is an essential research issue in data science and lays the foundation for realizing machine intelligence.

In this book, we discuss how to unleash hidden knowledge buried in unstructured text. We propose to first structure raw text using taxonomies and then analyze structured text in a more fine-grained and semantic way. Due to the diversity of application scenarios, different corpora or different use cases may call for different taxonomies. For example, one analyst aiming to find experts in different scientific areas may want a field-of-study taxonomy, while another analyst who studies the technology readiness may call for a taxonomy capturing technology dependencies. Moreover, even within one taxonomy, we also enable users to organize concepts at their will, such as with different levels containing concepts of different categories. For instance, in a computer science taxonomy, top levels could be about *field of studies*, intermediate levels may discuss *research tasks*, and the bottom levels can cover *evaluation metrics*. Asking human experts to manually curate those taxonomies, one for every possible application, is time-consuming, costly, and unscalable. Therefore, we propose to automatically discover and explore taxonomies based on the datasets and applications, with critical but minimal human guidance.

This book outlines a data-driven approach that automatically constructs, enriches, and applies taxonomies for unleashing knowledge from massive unstructured text. Particularly, we investigate four areas of research, including:

1. **Concept Set Discovery.** To obtain concept nodes in the taxonomy, we first develop a collection of concept set expansion methods to extract concepts from text corpora by expanding a small set of seed concepts into a complete list of concepts that belong to the same semantic class.
2. **Taxonomy Construction.** To organize above identified concepts into hierarchical structure, we propose a set of taxonomy construction methods to discover taxonomic relations among concepts by analyzing example relation instances (i.e. , concept pairs

indicating the target relation semantics) and utilizing distant supervision from existing, open-domain knowledge bases.

3. **Taxonomy Enrichment.** As human knowledge is constantly growing, a static taxonomy may fail to capture emerging user needs. Thus, a taxonomy enrichment step would be essential to keep our taxonomies up-to-date in real-world applications. We facilitate this process by expanding the taxonomy to incorporate new concepts.

4. **Taxonomy-Guided Classification.** After an up-to-date taxonomy is obtained, we develop principled methods to leverage taxonomies for classification tasks.

Together, these pieces constitute an integrated framework for leveraging taxonomies to convert massive text data into actionable knowledge.

New York City, USA   Jiaming Shen

Contents

Introduction

1.1 Overview

Born in an era of information explosion, we are inevitably inundated with vast amounts of text data. Every day, there are thousands of scientific papers, tens of thousands of corporate reports, product reviews, and millions of social media posts produced and shared worldwide. Their volume itself is massive, and more importantly, it keeps growing. When properly analyzed, these text data can be game-changing for science, engineering, business intelligence, policy design, e-commerce, and more. Consequently, turning those massive text data into actionable knowledge is an essential research issue in data science and lays the foundation for realizing machine intelligence.

With massive unstructured text stored or streaming in dynamically, we realize an important methodology for turning them into knowledge is to first 'structure' them. Instead of working on unstructured raw text directly, we can first utilize a taxonomy to structure them and then analyze structured text corpora in a more fine-grained and semantic way. This strategy can accelerate the knowledge discovery process and enable machines to digest the knowledge in an efficient and effective way (Fig. 1.1).

There are many existing taxonomies in different domains (e.g., MeSH [5], ACM CCS [1], Pinterest Taxonomy [3], etc.). Most of these taxonomies, however, are curated by human exports, which is costly, time-consuming, and non-scalable. More importantly, even with massive human labeling efforts, we could eventually only obtain one taxonomy but a single taxonomy cannot fit all applications. For example, given the same set of computer science literature, one user may want to identify experts in different fields, and thus calls for a field-of-studies taxonomy, while another person may want to analyze the readiness of different technologies, and thus needs a taxonomy capturing technology dependencies. Naturally, we need to build taxonomies based on the corpora to be analyzed and the applications to be explored. Sometimes, even for the same application, different domain experts may have their own points of view and thus we should ideally enable them to organize concepts at their

J. Shen and J. Han, *Automated Taxonomy Discovery and Exploration*,
Synthesis Lectures on Data Mining and Knowledge Discovery,
https://doi.org/10.1007/978-3-031-11405-2_1

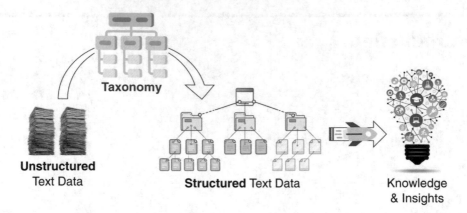

Fig. 1.1 Turning unstructured text data to knowledge and insights by utilizing a taxonomy to structure raw text

will, such as with different levels including concepts of different categories. For example, in the computer science domain, top levels could be about *field of studies* (e.g., data mining, natural language processing, etc.), immediate levels may discuss *research tasks* (e.g., outlier detection, machine translation, etc.), and the bottom levels cover *evaluation metrics* (e.g., F1 score, NDCG, BLUE, etc.). As a result, it is unrealistic to ask humans to create a taxonomy for each application.

In this book, we aim to develop principled methods to automatically construct, enrich, and explore taxonomies for knowledge discovery from text data (Fig. 1.2). Our methods alleviate the need for heavy human annotations by utilizing *distant supervision* from existing, open knowledge bases, *weak supervision* from a few user-provided examples, and *self*

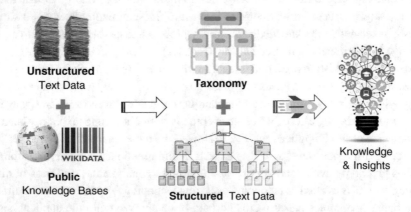

Fig. 1.2 Utilizing automatically constructed taxonomy for converting text data into actionable knowledge and insights

supervision from signals in massive unlabeled data. Particularly, we investigate four areas of studies, including: (1) **Concept Set Discovery**. To obtain concept nodes in the taxonomy, we first develop a collection of concept set expansion methods [10, 22, 23] to extract concepts from text corpora by expanding a small set of seed concepts into a complete list of concepts that belong to the same semantic class; (2) **Taxonomy Construction**. To organize above identified concepts into hierarchical structure, we propose a set of taxonomy construction methods [11, 14] to discover taxonomic relations among concepts by analyzing example relation instances (i.e., concept pairs indicating the target relation semantics) and utilizing distant supervision from existing, open-domain knowledge bases; (3) **Taxonomy Enrichment**. As human knowledge is constantly growing, a static taxonomy may fail to capture emerging user needs. Thus, a taxonomy enrichment step would be essential to keep our taxonomies up-to-date in real-world applications. We facilitate this process by expanding the taxonomy to incorporate new concepts [9, 21]. (4) **Taxonomy-Guided Classification**. After an up-to-date taxonomy is obtained, we develop principled methods to leverage taxonomies for classification tasks [6–8]. Together, these pieces constitute an integrated framework for leveraging taxonomies to convert massive text data into actionable knowledge, as shown in Fig. 1.3.

Finally, we want to emphasize that although the taxonomy structure discussed in this book shares some commonalities with knowledge graph [2], these two structured data formats have different focuses. Taxonomies capture more abstract concept-level ideas (e.g., one research area, one type of diseases, etc.) while knowledge graphs contain more information about concrete entity-level things (e.g., a celebrity, a famous institution, etc.). In many cases, taxonomies and knowledge graphs contain complementary world knowledge and they have the potentials to mutually enhance each other.

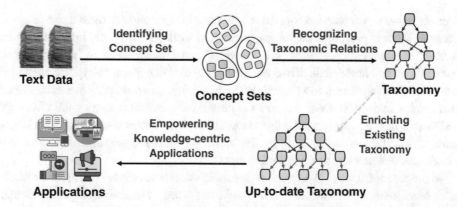

Fig. 1.3 An integrated framework for leveraging the taxonomy to unleash hidden knowledge buried in massive unstructured text

1.2   Technical Roadmap

We present a concrete roadmap for leverage taxonomies to turn unstructured text into structured knowledge below.

1.2.1   Concept Set Expansion

The first step towards building a corpus- and application-specific concept taxonomy is to
identify a set of user-interested concepts from the given corpus. We accomplish this goal
by proposing SetExpan [10], a concept set expansion method that automatically expands
a small set of seed concepts into a completed list of concepts belong to the same user-
interested category. SetExpan is an iterative algorithm and includes two core techniques.
The first one is a context feature selection method that chooses clean context features for
calculating concept-concept distributional similarity, and The second technique is a ranking-
based unsupervised ensemble method for expanding the concept set based on selected context
features. Experiments have shown that SetExpan can select high-quality and interpretable
context features and outperforms previous best methods by more than 32.6% for concept
identification accuracy. Recently, we have further enhanced SetExpan in two ways: (1)
leveraging multiple negative sets to guard each other and avoid semantic drifting [4], and
(2) incorporating information from pre-trained language models to compensate for the weak
supervision signals from the seed set [22].

1.2.2   Taxonomy Construction

After obtaining user-interested concept sets, we continue to organize them into a taxonomy
structure. Most of previous taxonomy construction work [16–18] build taxonomies based
on "is-A" relations (e.g., a "*panda*" is a "*mammal*" and a "*mammal*" is an "*animal*") by first
leveraging pattern-based or distributional methods to extract hypernym-hyponym term pairs
and then organizing them into a tree-structured hierarchy. However, such hierarchies cannot
satisfy many real-world needs due to its (1) *inflexible semantics*: many applications may
need hierarchies carrying more flexible semantics such as "*city-state-country*" in a location
taxonomy; and (2) *limited applicability*: the "universal" taxonomy so constructed is unlikely
to fit diverse and user-specific application tasks.

   We propose HiExpan [11], the first *task-guided* concept taxonomy construction method
which takes a user-provided "seed" taxonomy tree (as task guidance) along with a domain-
specific corpus and generates a desired taxonomy automatically. For example, a user may
provide a seed taxonomy containing only two countries and two states along with a large
corpus, and HiExpan will output a taxonomy which covers all the countries and states mentioned in the corpus. HiExpan captures weak supervision signals in the seed taxonomy and

iteratively expands the seed taxonomy into a fully-fledged taxonomy in a top-down hier-archical way. Specifically, HiExpan views all children under each taxonomy node forming a coherent set and builds the taxonomy by recursively expanding all these sets. Further-more, HiExpan incorporates a weakly-supervised relation extraction module to extract the initial children of a newly-expanded node and adjusts the taxonomy tree by optimizing its global structure. Experiments have demonstrated the effectiveness of HiExpan for build-ing meaningful taxonomies in various domains, including news, computer science, and life science.

1.2.3 Taxonomy Enrichment

As human knowledge is constantly growing, it is necessary to expand or enrich an existing concept taxonomy to incorporate new knowledge and be adapted to real-world applications. One naive solution is to re-run the entire taxonomy construction process from scratch. Although being intuitive, this approach has several limitations. First, many taxonomies have a top-level design provided by domain experts and such design shall be preserved. Second, a newly constructed taxonomy may not be consistent with the old one, which can lead to instabilities of its dependent downstream applications. Finally, as targeting the scenario of building taxonomy from scratch, most previous methods are unsupervised and cannot leverage signals from the existing taxonomy to construct a new one.

We propose TaxoExpan [9] to tackles taxonomy expansion problem—the process of automatically incorporating new concepts into an existing taxonomy. A key challenge for concept taxonomy expansion is the lack of labeled data. TaxoExpan addresses this problem by generating a set of ⟨query concept, anchor concept⟩ pairs from the existing taxonomy as self-supervision data. TaxoExpan first uses a position-enhanced graph neural network to encode each anchor concept's local structure in the existing taxonomy. Then, it learns to predict whether a query concept is the direct child of an anchor concept in the taxonomy using a noise-robust training objective. TaxoExpan can successfully expand a large field-of-study taxonomy with hundreds of thousands of concepts and outperforms the winning solution of SemEval 2016 taxonomy expansion task by 6.8% while running orders of magnitude faster. Recently, we have further improved TaxoExpan by: (1) deriving concept mini-paths from the existing taxonomy as self-supervision data while learning the model using multi-view co-training [19], and (2) identifying both the parent and children for each new emerging concept in the existing taxonomy [15, 21].

1.2.4 Taxonomy-Guided Classification

With a concept taxonomy constructed and enriched on a domain-specific document col-lection, we can explore a lot of downstream applications. For example, we have utilized taxonomies to facilitate semantic literature search [12, 13, 20] or to empower job post recommendation [14]. One important prerequisite of all those applications is that the text unit (either an entire document or an in-context text span) need to be tagged with a set of classes in the corresponding taxonomy. This can be formulated as a hierarchical multi-label text classification (HMTC) problem. Most existing HMTC methods are supervised and require massive human labeled training data that are not available in many real world scenarios.

To fully exploit the power of taxonomy, we propose TaxoClass [8], a weakly-supervised framework using only class surface names for hierarchical multi-label text classification. TaxoClass alleviates heavy human-labeling burdens and thus has a broader application scope. Specifically, TaxoClass leverages the explicit class relations in the given class taxonomy and pinpoints a few most essential classes for each document as its "core" classes. Based on those core classes, TaxoClass first trains a taxonomy-enhanced classifier and then generalizes this classifier via multi-label self-training. Our experiments have shown TaxoClass can achieve around 0.71 Example-F1 using only class names, outperforming the state-of-the-art weakly-supervised methods by 25%.

1.3 Organization

The remainder of the book is organized as follows. We first discuss how to mine concept sets in Chap. 2 and how to construct a concept taxonomy from raw text corpora in Chap. 3. Then, we present our taxonomy enrichment technique in Chap. 4 and taxonomy-guided classification methods in Chap. 5. Finally, in Chap. 6, we conclude this book and describe our future work.

References

1. Cassel, L.N., Palivela, S., Marepalli, S., Padyala, A., Deep, R., Terala, S.: The new ACM CCS and a computing ontology. In: Proceedings of the 13th ACM/IEEE-CS Joint Conference on Digital Libraries (2013)
2. Dong, X., Gabrilovich, E., Heitz, G., Horn, W., Lao, N., Murphy, K., Strohmann, T., Sun, S., Zhang, W.: Knowledge vault: a web-scale approach to probabilistic knowledge fusion. In: Proceedings of the 20th ACM SIGKDD International Conference on Knowledge Discovery And Data Mining (2014)
3. Gonçalves, R.S., Horridge, M., Li, R., Liu, Y., Musen, M.A., Nyulas, C.I., Obamos, E., Shrouty, D., Temple, D.: Use of OWL and semantic web technologies at Pinterest. In: Proceedings of 2019 International Semantic Web Conference (2019)
4. Huang, J., Xie, Y., Meng, Y., Shen, J., Zhang, Y., Han, J.: Guiding corpus-based set expansion by auxiliary sets generation and co-expansion. In: Proceedings of the 2020 Web Conference (2020)
5. Lipscomb, C.E.: Medical subject headings (MeSH). In: Bulletin of the Medical Library Association (2000)

6. Meng, Y., Shen, J., Zhang, C., Han, J.: Weakly-supervised neural text classification. In: Proceedings of the 27th ACM International Conference on Information and Knowledge Management (2018)

7. Meng, Y., Shen, J., Zhang, C., Han, J.: Weakly-supervised hierarchical text classification. In: Proceedings of the 2019 AAAI Conference on Artificial Intelligence (2019)

8. Shen, J., Qiu, W., Meng, Y., Shang, J., Ren, X., Han, J.: TaxoClass: hierarchical multi-label text classification using only class names. In: Proceedings of the 2021 Conference of the North American Chapter of the Association for Computational Linguistics: Human Language Technologies (2021)

9. Shen, J., Shen, Z., Xiong, C., Wang, C., Wang, K., Han, J.: TaxoExpan: self-supervised taxonomy expansion with position-enhanced graph neural network. In: Proceedings of the 2020 Web Conference (2020)

10. Shen, J., Wu, Z., Lei, D., Shang, J., Ren, X., Han, J.: SetExpan: corpus-based set expansion via context feature selection and rank ensemble. In: Proceedings of the 2017 Joint European Conference on Machine Learning and Knowledge Discovery in Databases (2017)

11. Shen, J., Wu, Z., Lei, D., Zhang, C., Ren, X., Vanni, M., Sadler, B.M., Han, J.: HiExpan: task-guided taxonomy construction by hierarchical tree expansion. In: Proceedings of the 24th ACM SIGKDD International Conference on Knowledge Discovery & Data Mining (2018)

12. Shen, J., Xiao, J., He, X., Shang, J., Sinha, S., Han, J.: Entity set search of scientific literature: an unsupervised ranking approach. In: Proccedings of the 41st International ACM SIGIR Conference on Research & Development in Information Retrieval (2018)

13. Shen, J., Xiao, J., Zhang, Y., Yang, C., Shang, J., Han, J., Sinha, S., Ping, P., Weinshilboum, R., Lu, Z., Han, J.: SetSearch+: entity-set-aware search and mining for scientific literature. In: Proceedings of the 24th ACM SIGKDD International Conference on Knowledge Discovery and Data Mining (Demo) (2018)

14. Shi, Y., Shen, J., Li, Y., Zhang, N., He, X., Lou, Z., Zhu, Q., Walker, M., Kim, M.H., Han, J.: Discovering hypernymy in text-rich heterogeneous information network by exploiting context granularity. In: Proceedings of the 28th ACM International Conference on Information and Knowledge Management (2019)

15. Song, X., Shen, J., Zhang, J., Han, J.: Who should go first? a self-supervised concept sorting model for improving taxonomy expansion. In: Proceedings of the International Workshop on Self-Supervised Learning for the Web (2021)

16. Velardi, P., Faralli, S., Navigli, R.: Ontolearn reloaded: a graph-based algorithm for taxonomy induction. In: Computational Linguistics (2013)

17. Wang, C., He, X., Zhou, A.: A short survey on taxonomy learning from text corpora: issues, resources and recent advances. In: Proceedings of the 2017 Conference on Empirical Methods in Natural Language Processing (2017)

18. Wu, W., Li, H., Wang, H., Zhu, K.Q.: Probase: a probabilistic taxonomy for text understanding. In: Proceedings of the 2012 ACM SIGMOD International Conference on Management of Data (2012)

19. Yu, Y., Li, Y., Shen, J., Feng, H., Sun, J., Zhang, C.: STEAM: self-supervised taxonomy expansion with mini-paths. In: Proceedings of the 26th ACM SIGKDD International Conference on Knowledge Discovery and Data Mining (2020)

20. Zha, H., Shen, J., Li, K., Greiff, W., Vanni, M., Han, J., Yan, X.: FTS: Faceted taxonomy construction and search for scientific publications. In: Proceedings of the 24th ACM SIGKDD International Conference on Knowledge Discovery and Data Mining (Demo) (2018)

21. Zhang, J., Song, X., Zeng, Y., Chen, J., Shen, J., Mao, Y., Li, L.: Taxonomy completion via triplet matching network. In: Proceedings of the 2021 AAAI Conference on Artificial Intelligence (2021)

22. Zhang, Y., Shen, J., Shang, J., Han, J.: Empower entity set expansion via language model probing. In: Proceedings of the 58th Annual Meeting of the Association for Computational Linguistics (2020)
23. Zhu, W., Gong, H., Shen, J., Zhang, C., Shang, J., Bhat, S., Han, J.: FUSE: multi-faceted set expansion by coherent clustering of skip-grams. In: Proceedings of 2020 Joint European Conference on Machine Learning and Knowledge Discovery in Databases (2020)

Concept Set Expansion

2

2.1 Overview and Motivations

Concept set expansion refers to the problem of expanding a small set of seed concepts into a complete set of concepts that belong to the same semantic class [34]. For example, if a given seed set is {*Oregon, Texas, Iowa*}, concept set expansion should return a hopefully complete set of concepts in the same semantic class, "*U.S. states*". Concept set expansion can benefit various downstream applications, such as knowledge extraction [8], taxonomy induction [32], and web search [2].

One line of work (e.g., *Google Set* [31] and *SEAL* [34]) solves this task by submitting a query consisting of seed concepts to a search engine and mining top-ranked webpages. While this approach can achieve relatively good quality, the required seed-oriented online data extraction is costly. Therefore, more studies [10, 20, 24, 27, 33] are proposed in a *corpus-based* setting where sets are expanded by offline processing based on a specific corpus.

There are two general approaches for *corpus-based* set expansion—*one-time concept ranking* and *iterative pattern-based bootstrapping*. Based on the assumption that similar concepts appear in similar contexts, the first approach [10, 20, 27] makes a one-time ranking of candidate concepts based on their distributional similarity with seed concepts. A variety of "contexts" are used, including Web table, Wikipedia list, or just free-text patterns, and concept-concept distributional similarity is calculated based on *all* context features. However, blindly using *all* such features can lead to *concept intrusion* errors. Namely, some undesired concepts are wrongly introduced into the expanded set because many context features are not representative for defining the target semantic class although they do have connections with some of the seed concepts.

The second approach, iterative pattern-based bootstrapping [8, 9, 26], starts from seed concepts to extract quality patterns, based on a predefined pattern scoring mechanism, and it

© The Author(s), under exclusive license to Springer Nature Switzerland AG 2022
J. Shen and J. Han, *Automated Taxonomy Discovery and Exploration*,
Synthesis Lectures on Data Mining and Knowledge Discovery,
https://doi.org/10.1007/978-3-031-11405-2_2

then applies extracted patterns to obtain even higher quality concepts using another concept scoring method. This process iterates and the high-quality patterns from all previous iterations are accumulated into a pattern pool which will be used for the next round of concept extraction. This approach works only when patterns/concepts extracted at each iteration are highly accurate, otherwise, it may cause severe *semantic shift* problem. Suppose in the previous example, "*located in __*" is taken as a good pattern from the seed set *{Oregon, Texas, Iowa}*, and this pattern brings in *USA* and *Ontario*. These undesired concepts may bring in even lower quality patterns and iteratively cause the set shifting farther away. Thus, the pattern and concept scoring methods are crucial but sensitive in iterative bootstrapping methods. If they are not defined perfectly, the semantic shift can cause big problems. However, it is hard to have a perfect scoring mechanism due to the diversity and noisiness of unstructured text data.

To address these challenges, we first propose a novel set expansion framework SetExpan in this chapter. SetExpan carefully and conservatively extracts each candidate concept and iteratively improves the results. First, to overcome the concept intrusion problem, instead of using all context features, context features are carefully selected by calculating distributional similarity. Second, to overcome the semantic drift problem, different from other bootstrapped approaches, our high-quality feature pool will be reset at the beginning of each iteration. Finally, our carefully designed unsupervised ranking-based ensemble method is used at each iteration to further refine concepts and make our system robust to noisy or wrongly extracted pattern features. Figure 2.1 shows the pipeline at each iteration. SetExpan iteratively expands a concept set through a context feature selection step and a concept selection step. At the context feature selection, each context feature is scored based on its strength with currently expanded concepts and top-ranked context features are selected. At the concept selection step, multiple subsets of the selected representative context features are sampled and each subset is used to obtain a ranked concept list. Finally, all the ranked lists are collected to compute the final ranking list of each candidate concept for expansion.

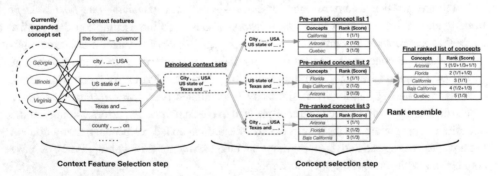

Fig. 2.1 An example showing two steps in one iteration of SetExpan

The major contributions of this chapter are highlighted as follows:

- We propose an iterative set expansion framework with a novel context feature selection approach, to handle the issues of concept intrusion and semantic drift.
- We develop an unsupervised ranking-based ensemble algorithm for concept selection to make our system robust and further reduce the impact of semantic drift.
- We demonstrate the effectiveness and efficiency of our methods and show improvements over prior methods on multiple real-world datasets in different domains (news, Wikipedia articles, and scientific papers).

The rest of this chapter is organized as follows. We first discuss some related work in Sect. 2.2. Then, we propose our SetExpan framework in Sect. 2.3 and present experiment results in Sect. 2.4. After that, we discuss how to extend SetExpan in Sect. 2.5. Finally, we conclude this chapter in Sect. 2.6.

2.2   Related Work

The problem of completing a concept set given several seed concepts has attracted extensive research efforts due to its practical importance. Google Sets [31] was among the earliest work dealing with this problem. It used proprietary algorithms and is no longer publicly accessible. Later, Wang and Cohen proposed *SEAL* system [34], which first submits a query consisting of all seed concepts into a general search engine and then mines the top-ranked webpages. Recently, Chen et al. [2] improved this approach by leveraging a "page-specific" extractor built in a supervised manner and showed good performance on long-tail (i.e., rare) term expansion. All these methods need an external search engine and require seed-oriented data extraction. In comparison, our approach conducts corpus-based set expansion without resorting to online data extraction from specific webpages.

To tackle the corpus-based set expansion problem, Ghahramani and Heller [6] used a Bayesian method to model the probability that a candidate concept belongs to some unknown cluster that contains the input seeds. Pantel et al. [20] developed a web-scale set expansion pipeline by exploiting distributional similarity on context words for each candidate concept. He et al. [10] proposed the SEISA system that uses query logs along with web lists as external evidence besides free text, and designed an iterative similarity aggregation function for set expansion. Recently, Wang et al. [33] leveraged web tables and showed very competitive results when not only seed concepts but also intended class name were given. While these semi-structured lists and tables are helpful, they are not always available for some specific domain corpus such as PubMed articles or DBLP papers. Perhaps the most relevant work to ours is by Rong [24]. In that paper, the authors used the skip-pattern feature combined with additional user-generated ontologies (i.e., Wikipedia list) for set expansion. However, they targeted the multifaceted expansion and exploited all skip-pattern features for calculating the similarity because two concepts. In our work, we keep the core idea of distributional similarity but calculate such similarity using only carefully selected *denoised* context features.

In a broader sense, our work is also related to information extraction and named concept recognition. Without given enough training data, bootstrapped concept extraction system [7, 8] is the most popular and effective choice. At each bootstrap iteration, the system will first create patterns around concepts; score patterns based on their ability to extract more positive concepts and less negative concepts (if provided), and use top-ranked patterns to extract more candidate concepts. Multiple pattern scoring and concept scoring functions are proposed. For example, Riloff et al. [23] scored each pattern by calculating the ratio of positive concepts among all concepts extracted by it, and scored each candidate concept by the number and quality of its matched patterns. Gupta et al. [7] scored patterns using the ratio of scaled frequencies of positive concepts among all concepts extracted by it. All these methods are heuristic and sensitive to different model parameters.

More generally, our work is also related to class label acquisition [28, 35] which aims to propagate class labels to data instances based on labeled training examples, and concept clustering [1, 13] where the goal is to find clusters of concepts. However, the class label acquisition methods require a much larger number of training examples than the typical size of user input seed set, and the concept clustering algorithms can only find semantically related concepts instead of concepts strictly in the same semantic class.

2.3   SetExpan: Weakly-Supervised Concept Set Expansion

We first introduce our context features and data model used by SetExpan in Sect. 2.3.1 and then present our context-dependent similarity measure in Sect. 2.3.2. After that, we discuss how to select context features in Sect. 2.3.3 and present our novel unsupervised ranking-based ensemble method for concept selection in Sect. 2.3.4.

2.3.1   Data Model and Context Features

We explore two types of context features obtained from the plain text: (1) skip-patterns [24] and (2) coarse-grained types [8]. As shown in Fig. 2.2a, data is modeled as a bipartite graph, with candidate concepts on one side and their context features on the other. Each type of context features are described as follows.

Skip-pattern: Given a target concept e_i in a sentence, one of its skip-pattern is "w_{-1} _ w_1" where w_{-1} and w_1 are two context words and e_i is replaced with a placeholder. For example, one skip-pattern of concept "*Illinois*" in sentence "*We need to pay Illinois sales tax.*" is "pay_sales". As suggested in [24], we extract up to six skip-patterns of different lengths for one target concept e_i in each sentence. One advantage of using skip-patterns is that it imposes strong positional constraints.

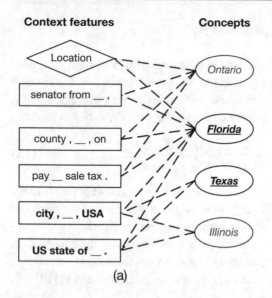

Fig. 2.2 a A simplified bipartite graph data model. **b** Similarity with seed concept conditioned on two different sets of context features

Coarse-grained type: Besides the unstructured skip-pattern features, we use coarse-grained types to filter those obviously-wrong concepts. For examples, when we expand the "U.S. states" class, we will not consider any concept that is typed as a person. After this process, we can obtain a cleaner subset of candidate concepts. Such a mechanism is also adopted in [8].

After obtaining the "nodes" in bipartite graph data model, we need to model the edges in the graph. In this work, we assign the weight between each pair of concept e and context feature c using the *TF-IDF transformation* [24], which is calculated as follows:

$$f_{e,c} = \log(1 + X_{e,c}) \left[\log |E| - \log \left(\sum_{e'} X_{e',c} \right) \right], \qquad (2.1)$$

where $X_{e,c}$ is the raw co-occurrence count between concept e and context feature c, $|E|$ is the total number of candidate concepts. We refer to such scaling as the *TF-IDF transformation* since it resembles the *tf-idf* scoring in information retrieval if we treat each concept e as a "document" and each of its context feature c as a "term". Empirically, we find such weight scaling outperforms some other alternatives such as point-wise mutual information (PMI) [10], truncated PMI [18], and BM25 scoring [22].

2.3.2   Context-Dependent Concept Similarity

With the bipartite graph data model constructed, the task of expanding a concept set at each iteration can be viewed as finding a set of concepts that are most "similar" to the currently expanded set. In this study, we use the weighted Jaccard similarity measure. Specifically, given a set of context features F, we calculate the *context-dependent* similarity as follows:

$$Sim(e_1, e_2 | F) = \frac{\sum_{c \in F} \min(f_{e_1,c}, f_{e_2,c})}{\sum_{c \in F} \max(f_{e_1,c}, f_{e_2,c})}. \tag{2.2}$$

Notice that if we change context feature set F, the similarity between concept pair is likely to change, as demonstrated in Fig. 2.2b. Finally, we want to emphasize that our proposed method is general in the sense that other common similarity metrics (e.g., cosine similarity) can also be used. In practice, we find the performance of a set expansion method depends less on the exact choice of base similarity metrics, but more on which contexts are selected for calculating *context-dependent* similarity. Similar results were also reported in [10].

2.3.3   Context Feature Selection

As shown in Fig. 2.2b, the similarity between two concepts really depends on the selected feature set F. The motivation of context feature selection is to find a feature subset F^* of fixed size Q that best "profiles" the target semantic class. In other words, we want to select a feature set F^* based on which concepts within target class are most "similar" to each other. Given such F^*, the concept-concept similarity conditioned on it can best reflect their distributional similarity with regard to the target class. In some sense, such F^* best profiles the target semantic class. Unfortunately, to find such F^* of fixed size Q, we need to solve the following optimization problem which turns out to be NP-Hard, as shown in [3].

$$F^* = \underset{|F|=Q}{\arg\max} \sum_{i=1}^{|X|} \sum_{j>i}^{|X|} Sim(e_i, e_j | F), \tag{2.3}$$

where X is the set of currently expanded concepts. Initially, we treat the user input seed set S as X. As iterations proceed, more concepts will be added into X.

Given the NP-hardness of finding the optimal context feature set, we resort to a heuristic method that first scores each context feature based on its accumulated strength with concepts in X and then selects top Q features with maximum scores. Take Fig. 2.2a as an example, we assume all edge weights in the bipartite graph are equal to 1 and let the currently expanded concept set X be "{Florida", "Texas"}. Suppose we want to select two "denoised" context features, we will first score each context feature based on its associated concepts in X. The top 4 contexts will obtain a score 1 since they match only one concept in X with strength 1, and the 2 contexts below will get a score 2 because they match both concepts in X. Then, we rank context features based on their scores and select 2 contexts with highest scores: "city, _, USA", "US state of _ ." into F.

Finally, we want to emphasize two major differences of our context feature selection method from other heuristic "pattern selection" methods. First, most pattern selection methods require either users to explicitly provide the "negative" examples for the target semantic class [8, 12, 26], or implicitly expand multiple mutually exclusive classes in which instances in one class serve as negative examples for all the other classes [4, 18]. Our method requires only a small number of "positive" examples. In most cases, it is hard for humans to find good discriminative negative examples for one class, or to provide both mutually exclusive and somehow related comparative classes. Second, the bootstrapping method will add its selected "quality patterns" during each iteration into a quality pattern pool, while our method will select high quality context features at each iteration from scratch. If one noisy pattern is selected and added into the pool, it will continue to introduce more irrelevant concepts at all the following iterations. Our method can avoid such noise accumulation.

2.3.4 Concept Selection via Rank Ensemble

Intuitively, the concept selection problem can be viewed as finding those concepts that are most similar to the currently expanded set X conditioned on the selected context feature set F. To achieve this, we can rank each candidate concept based on its score in Eq. (2.4) and then add top-ranked ones into the expanded set:

$$score(e|X, F) = \frac{1}{|X|} \sum_{e' \in X} Sim(e, e'|F). \tag{2.4}$$

However, due to the ambiguity of natural language in free-text corpora, the selected context feature set F may still be noisy in the sense that an irrelevant concept is ranked higher than a relevant one. To further reduce such errors, we propose a novel ranking-based ensemble method for concept selection.

The key insight of our method is that an inferior concept will not appear frequently in multiple pre-ranked concept lists at top positions. Given a selected context set F, we first use sampling without replacement method to generate T subsets of context features $F_t, t = 1, 2, \ldots, T$. Each subset is of size $\alpha|F|$ where α is a model parameter within range $[0, 1]$. For each F_t, we can obtain a pre-ranked list of candidate concepts L_t based on $score(e|X, F_t)$ defined in Eq. (2.4). We use r_t^i to denote the rank of concept e_i in list L_t. If concept e_i does not appear in L_t, we let $r_t^i = \infty$. Finally, we collect T pre-ranked lists and score each concept based on its mean reciprocal rank (mrr). All concepts with average rank above r, namely $mrr(e) \leq T/r$, will be added into concept set X.

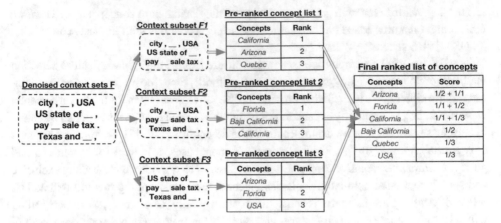

Fig. 2.3 A toy example to show concept selection via rank ensemble

---

Algorithm 2.1: SetExpan

Input: Candidate concept set E, initial seed set S, concept-context graph G, expected size of
      output set K, model parameters $\{Q, T, \alpha, r\}$.

Output: The expanded set X.

1   $X = S$;

2   **while** $|X| \leq K$ **do**

3   Set $F = \emptyset$ // Select denoised contexts from scratch;

4   Score context features based on X and add top Q denoised contexts into F;

5   // concept-selection via rank ensemble;

6   **for** $t = 1, 2, \ldots, T$ **do**

7   Uniformly sample αQ contexts and construct feature subset F_t;

8   Score concepts based on Eq. (2.4) given F_t and obtain the pre-ranked list L_t;

9   Update the mrr score of each concept based on Eq. (2.5);

10     $X = X \cup \{e | mrr(e) \geq \frac{T}{r}\}$ // Add concepts into expanded set X;

11 Return X;

$$mrr(e_i) = \sum_{t=1}^{T} \frac{1}{r_t^i}, \quad r_t^i = \sum_{e_j \in E} I\left(score(e_i|X, F_t) \leq score(e_j|X, F_t)\right), \quad (2.5)$$

where $I(\cdot)$ is the indicator function. Naturally, a relevant concept will rank at top position in multiple pre-ranked lists and thus accumulate a high mrr score, while an irrelevant concept will not consistently appear in multiple lists at high position which leads to low mrr score.

We use the following example to demonstrate the whole process of concept selection. In Fig. 2.3, we want to expand the "US states" semantic class given a selected context feature set F with 4 features. We first sample a subset of 3 context features $F_1 =$ "{city , __ , USA", "US state of __ ,", "pay __ sales tax .}", and then use F_1 to obtain a pre-ranked concept list

$L_1 = \langle$"California", "Arizona", "Quebec"$\rangle$. By repeating this process three times, we get three pre-ranked lists and ensemble them into a final ranked list in which concept "Arizona" is scored 1.5 because it is ranked in the 2nd position in L_1 and 1st position in L_3. Finally, we add those concepts with mrr score larger than 1, meaning this concept is ranked at 3rd position on average, into the expanded set X. In this simple example, the model parameters $T = 3$, $\alpha = \frac{|F_1|}{|F|} = 0.75$, and $r = 3$.

Summary. Algorithm 2.1 summarizes the whole SetExpan framework. The candidate concept set E and bipartite graph data model G are pre-calculated and stored. A user needs only to specify the seed set S as the task guidance and the expected size of output set K. There is a total of 4 model parameters: the number of top quality context features selected in each iteration Q, the number of pre-ranked concept lists T, the relative size of feature subset $0 < \alpha < 1$, and final mrr threshold r. The tuning and sensitivity of these parameters will be discussed in the experiment section.

2.4 Experiments

In this section, we will evaluate SetExpan on three massive text corpora across different domains. We first compare the propose method with many other methods to demonstrate its high performance. Then, we explore the robustness of our method by varying different hyper-parameters and present some interesting case studies.

2.4.1 Datasets

We use three corpora to evaluate the performance of SetExpan. Table 2.1 lists 3 datasets we used in experiments: (1) **APR** is constructed by crawling all 2015 news articles from AP and Reuters; (2) **Wiki** is a subset of English Wikipedia used in [15], and (3) **PubMed-CVD** is a collection of research paper abstracts about cardiovascular disease retrieved from PubMed. For APR and PubMed-CVD datasets, we adopt a data-driven phrase mining tool [16] to obtain concept mentions and type them using ClusType [21]. Each concept mention is mapped heuristically to a concept based on its lemmatized surface name. We then extract variable-length skip-patterns for all concept mentions as features for their corresponding concepts, and construct the bipartite graph data model as introduced in the previous section. For Wiki dataset, the concepts have already been extracted and typed using distant supervision. For the type information in each dataset, there are 16 coarse-grained types in APR and 4 coarse-grained types in PubMed-CVD. For Wiki, since it originally has about 50 fine-grained types, which may reveal too much information, we manually mapped them to 11 more coarse-grained types.

A query is a set of seed concepts of the same semantic class in a dataset, serving as the input for each system to expand the set. The process of query generation is as follows. For

Table 2.1 Datasets statistics and query descriptions

Dataset	FileSize	# Sentences (M)	# Concepts (K)	# Test queries
APR	775 MB	1.01	122	40
Wiki	1.02 GB	1.50	710	20
PubMed-CVD	9.3 GB	23	179	5

each dataset, we first extract 2000 most frequent concepts in it and construct a concept list. Then, we ask three volunteers to manually scan the concept lists and propose a few semantic classes for each list. The proposed class should be interesting, relatively unambiguous and has a reasonable coverage in its corresponding corpus. These semantic classes cover a wide variety of topics, including locations, companies as well as political parties, and have different degrees of difficulty for set expansion. After finalizing the semantic classes for each dataset, the volunteers randomly select concepts of each semantic class from the frequent concept list to form 5 queries of size 3. To select the queries for PubMed-CVD, we seek help from two additional volunteers with biomedical expertise, following the same previous approach. Due to the large size of PubMed-CVD dataset and runtime limitation, we only select 1 semantic class (hormones) with 5 queries.

With all queries selected, we have humans to label all the classes and instances returned by each of the following compared methods. For APR and Wiki datasets, the inter-rater agreements (kappa-value) over three students are 0.7608 and 0.7746, respectively. For PubMed-CVD dataset, the kappa-value is 0.9236. All concepts with conflicting label results are further resolved after discussions among all human labelers.

2.4.2 Compared Methods

Since the focus on this work is the corpus-based set expansion, we do not compare with other methods that require online data extractions. Also, to further analyze the effectiveness of each module in SetExpan framework. We implement 3 variations of our framework.

- word2vec [19]: We use the SkipGram model in word2vec to learn concept embeddings and return k nearest neighbors around seed concepts as the expanded set.
- PTE [29]: We first construct a heterogeneous information network including concepts, skip-pattern features, and type features. Then, we apply PTE model to learn concept embeddings and determine the k nearest neighbors around seed concepts.
- SEISA [10]: A concept set expansion algorithm based on iterative similarity aggregation. It uses the occurrence of concepts in web list and query log as concept features. In our

experiments, we replace the web list and query log with our skip-pattern and coarse-grained context features.

- EgoSet [24]: A multifaceted set expansion system based on skip-patterns, word2vec embeddings and WikiList. The original system expands a seed set to multiple concept sets, considering the ambiguities in seed set. To achieve this, they use a community detection method to separate extracted concepts into several groups. However, in order to better compare with EgoSet, we carefully select queries that have little ambiguity or at least the seed set in the query is dominating in one semantic class. Thus, we discard the community detection part in EgoSet and treat all extracted concepts as in one semantic class.

- SetExpan-cs: Disable the context feature selection module in SetExpan, and use all context features to calculate distributional similarity.

- SetExpan-re: Disable the rank ensemble module in SetExpan. Instead, we use all selected context feature to rank candidate concepts at one time and add top-ranked ones into the expanded set.

- SetExpan-full: The full version of our proposed method, with both context feature selection and rank ensemble components enabled.

For fair comparison, we try different combinations of parameters and report the best performance for each baseline method.

2.4.3 Evaluation Metrics

For each test case, the input is a query, which is a set of 3 seed concepts of the same semantic class. The output will be a ranked list of concepts. For each query, we use the conventional average precision $AP_k(c, r)$ at k ($k = 10, 20, 50$) for evaluation, given a ranked list of concepts c and an unordered ground-truth set r. For all queries under a semantic class, we calculate the mean average precision (MAP) at k as $\frac{1}{N} \sum_i AP_k(c_i, r)$, where N is the number of queries. To evaluate the performance of each approach on a specific dataset, we calculate the mean-MAP (MMAP) at k over all queried semantic classes as $MMAP_k = \frac{1}{T} \sum_{t=1}^{T} [(\frac{1}{N_t}) \sum_i AP_k(c_{ti}, r_t)]$, where T is the number of semantic classes, N_t is the number of queries of t-th semantic class, c_{ti} is the extracted concept list for i-th query for t-th semantic class, and r_t is the ground truth set for t-th semantic class.

2.4.4 Overall Performance

Table 2.2 shows the MMAP scores of all methods on 3 datasets.[1] We can see that SetExpan outperforms all four baselines in terms of the MMAP score. We further look at their

---

[1] Results of SEISA on PubMed-CVD are omitted due to the scalability issue.

Table 2.2 Set expansion performance on 3 datasets over all queries

Methods	APR			Wiki			PubMed-CVD		
	MAP@10	MAP@20	MAP@50	MAP@10	MAP@20	MAP@50	MAP@10	MAP@20	MAP@50
EgoSet	0.3949	0.3942	0.3706	0.5899	0.5754	0.5622	0.0511	0.0410	0.0441
SEISA	0.7423	0.6090	0.3892	0.7643	0.6606	0.4998	–	–	–
word2vec	0.6054	0.5385	0.4180	0.7193	0.6289	0.4510	0.8427	0.7701	0.6895
PTE	0.3144	0.2777	0.1996	0.6817	0.5596	0.3839	0.9071	0.7654	0.5641
SetExpan-cs	0.8240	0.7997	0.7674	0.9540	0.8955	0.7439	**1.000**	**1.000**	0.5991
SetExpan-re	0.8509	0.7792	0.7681	0.9392	0.8680	0.7291	**1.000**	0.9605	0.7371
SetExpan-full	**0.8967**	**0.8621**	**0.7885**	**0.9571**	**0.9010**	**0.7457**	**1.000**	**1.000**	**0.7454**

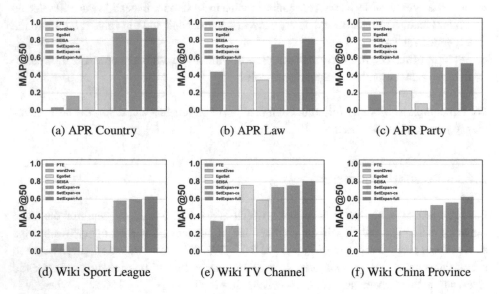

(a) APR Country (b) APR Law (c) APR Party

(d) Wiki Sport League (e) Wiki TV Channel (f) Wiki China Province

Fig. 2.4 Set expansion performance for each semantic class

performances on each concept class, as shown in Fig. 2.4 and each individual query, as shown in Fig. 2.5. We can see that the performance of these baseline methods varies a lot on different semantic classes, while our SetExpan can consistently beat them. One reason is that none of these methods applies context feature selection or rank ensemble, and a single set of unpruned features can lead to various levels of noise in the results. Another reason is the lack of an iterative mechanism in some of those approaches. For example, even if EgoSet includes the results from word2vec to help it boost the performance, it still achieves low MAP scores in some semantic classes. Finding the nearest neighbors in only one iteration can be a key reason. And although SEISA is applying the iterative technique, instead of adding a small number of new concepts in each iteration, it expands a full set in each iteration

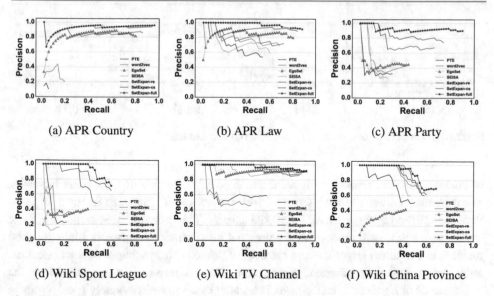

Fig. 2.5 Set expansion performance for individual queries in different classes

based on the coherence score of each candidate concept with the previously expanded set. It pre-calculates the size of the expanded set with the assumption that the feature similarities follow a certain distribution, which does not always hold to all datasets or semantic classes. Thus, if the size is far different from the actual size or is too big to extract a confident set at once, each iteration will introduce a lot of noise and cause semantic drift.

2.4.5 Ablation Studies

Comparison with SetExpan-re **and** SetExpan-cs. At the dataset level, the MMAP scores of SetExpan-full outperforms its two variation approaches. In the semantic class level, we can see that SetExpan-re and SetExpan-cs sometimes have their MAP much lower than SetExpan-full while sometimes they almost achieve the same performance with SetExpan-full. This means they fail to stably extract concepts with good quality. The main reason is still that a single set of features or ensembles over unpruned features can lead to various levels of noise in the results. Only under the circumstances that the single set of features or the unpruned features happen to be nicely selected without too much noise, which tends to happen when the query is relatively "easy", these variation approaches can achieve good results.

Effects of Context Feature Selection. We already see that adding the context feature selection component helps improve the performance. What's also noticeable is that the addition

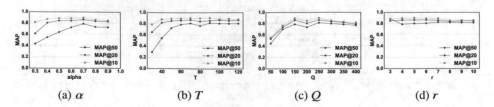

(a) α (b) T (c) Q (d) r

Fig. 2.6 Hyper-parameter sensitivity of SetExpan on two datasets

of context selection process becomes more obvious as the size of the corpus increases. The difference between MMAP scores of SetExpan-cs and SetExpan-full is much larger in PubMed-CVD compared with APR and Wiki datasets. This is because that as the corpus size increases, we will have more noisy features and more candidate concepts while the good features to define the target concept set may be limited. Thus, without context selection, noise can damage the performance much more. The evidence can also be found from the performance of EgoSet across the three datasets. It can achieve reasonably good results in APR and Wiki, however, it performs much worse in PubMed-CVD.

Effect of Rank Ensemble. From the above experiments, we can see that the effect of rank ensemble may vary across different semantic classes. However, the contribution of rank ensemble seems to be more stable across datasets, compared with the effect of context selection. This is because we apply the default set of hyper-parameter values in each test case above. In the below hyper-parameter analysis, we will show that the number of ensemble batches and the percentage of features to be randomly sampled can affect the contribution of rank ensemble to the concept set expansion performance.

Hyper-parameter Analysis. There are totally 4 hyper-parameters in SetExpan: Q (the number of selected context features), α (the percentage of features to be sampled), T (the number of ensemble batches), and r (the threshold of a candidate concept's average rank). We study the influence of each hyper-parameter by fixing all other hyper-parameters to default values, and present one graph showing the MMAP scores of SetExpan on APR dataset versus the changes of that hyper-parameter (Fig. 2.6).

- α: From the graph, the performance increases sharply as α increases until it reaches about 0.6. Then, it starts to stay stable and decreases after 0.7.
- T: The performance first increases as we increase the ensemble batches and then becomes stable after 60 batches.
- Q: In the range of 50–150, the performance increases sharply as Q increases, which means the majority of top 150 context features can provide rich information to identify concepts belonging to the target semantic class. The available information gets more and more saturated after Q reaches 150 and start to introduce noises and hamper the performance after around 300.

- r: Our experiments show that the performance is not very sensitive to the threshold of a candidate concept's average rank.

2.4.6 Case Studies

Figure 2.7 presents three case studies for SetExpan. We show one query for each dataset. In each case, we show top 3 ranked concepts and top/bottom 3 skip-pattern features after context feature selection for the first 3 iterations as well as the coarse-grained type. In all cases, our algorithm successfully extracts correct concepts in each iteration, and the top-ranked skip-patterns are representative in defining the target semantic class. On the other hand, we notice that most of the bottom 3 skip-patterns selected are very general or not representative at all. These context features could potentially introduce noisy concepts and thus the rank ensemble can play a rival role in improving the results.

Dataset	Query	Top ranked concepts in the first 3 iterations	Top/Bottom skip-pattern features selected in the first 3 iterations	Coarse-grained type
APR	{Patriot Act, Obamacare, Clery Act}	Iteration 1: USA Patriot Act, USA Freedom Act, Voting Rights Act, … Iteration 2: Stock Act, Religious Freedom Restoration Act, Foreign Intelligence Surveillance Act, … Iteration 3: Americans with Disabilities Act, Healthy Families Act, Goonda Act, …	Iteration 1: Top 3: "the __ provisions", "provisions of the __", "defund __." Bottom 3: "2010 __ ,", "also known as __ .", "under the __ , and" Iteration 2: Top 3: "under the __ to", "provisions of the __ ,", "the __ into law." Bottom 3: "the __ - which has", "the _The House", "the _ , first" Iteration 3: Top 3: "under the __ to", "Under the __ ,", "the __ into law" Bottom 3: "of the __ passed", "the __ , the most", "replacing __ ."	Event
Wiki	{ESPN, ESPN2, Spike TV}	Iteration 1: ABC, CBS, NBC, … Iteration 2: BBC, ITV, Channel 4, … Iteration 3: TBS, ITV1, BBC Two, …	Iteration 1: Top 3: "telecast on __ .", "televised on __ .", "televised by __ ." Bottom 3: "on __ , to the", ", and perhaps __ .", "from an __ website" Iteration 2: Top 3: "the __ sitcom", "the __ television network", "ABC , __ ," Bottom 3: "on __ on September", "broadcast on __ on", "the __ soap opera The" Iteration 3: Top 3: "the __ sitcom", "the __ soap", "the __ soap opera", … Bottom 3: "aired on __ between", "of the __ show", "on the __ crime"	Organization
PubMed -CVD	{FSH, TSH, MSH}	Iteration 1: LH, GH, ACTH, … Iteration 2: LHRH, AMH, GHRH, … Iteration 3: Renin, GnRH-I, AVP, …	Iteration 1: Top 3: "stimulating hormone (__)", "hormone (__) ,", "hormone (__) and", … Bottom 3: "g/L , __ =", ", __ and prolactin", "hormone (__) -" Iteration 2: Top 3: "hormone (__) ,", "hormone (__) and", "hormone (__) .", … Bottom 3: ", __ , estradiol ,", ", __ , and PRL", "hormone (__) -" Iteration 3: Top 3: "hormone (__) ,", "hormone (__) and", "hormone (__) .", … Bottom 3: "(__) and insulin-like", ", TSH , __ ,", "levels of __ , FSH"	Proteins and Genes (PRGE)

Fig. 2.7 Three case studies of SetExpan on each dataset

2.5   Extensions of SetExpan

SetExpan demonstrates an effective iterative framework for concept set expansion. Together with collaborators in our group, we further extend SetExpan by exploiting automatically discovered negative sets [11] and incorporating pre-trained language models [39].

2.5.1   Addressing Concept Drifts via Auxiliary Sets Generation and Co-expansion

We observe that a typical source of SetExpan error comes from the concepts from different semantic classes that share some common relations to the target class. For example, when expanding the *Country* class, we may wrongly introduce those erroneous concepts in the *City* class. If we can capture such subtle relationships between concepts belonging to different semantic classes, we can use them to separate concepts from different classes and conduct set co-expansion. Such co-expansion may incorporate signals from all the related, participating classes, keep warning the target class not to cross over the boundaries of its possible rivals, and guide the expanding direction of each set by avoiding to bump into each other's territory. The co-expansion of such multiple rival sets benefits each other from mutually exclusive signals, and the quality of multiple sets expansion can be improved simultaneously. Some previous studies [12, 14, 30] also found that using mutual exclusive signals from other related "auxiliary" classes could help. However, they often require users to explicitly provide those auxiliary classes, which was not applicable in many real-world scenarios.

We propose SetCoExpan, a fully automated approach to improve SetExpan without human-provided auxiliary sets. As shown in Fig. 2.8, SetCoExpan consists of two modules that are operated iteratively: (1) an auxiliary sets generation module that finds auxiliary sets holding certain relations with the target set in an unsupervised way, and (2) a multiple sets co-expansion module that takes multiple sets as input and extracts the most discriminative features to tell the target class from auxiliary sets. The auxiliary sets generation module first retrieves semantically related terms to each seed element in an embedding space that captures topical similarity. These related terms are then grouped by their semantic types, captured by

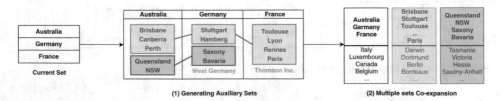

Fig. 2.8 Overview of SetCoExpan. For user-input country names, we first retrieve related terms such as provinces and cities and then cluster them into auxiliary sets. Multiple sets are then co-expanded by extracting discriminative context features

intra-seed clustering and inter-seed merging in an unsupervised way. The multiple sets co-expansion modules takes the target seed set as well as auxiliary sets as input. By incorporating knowledge from both seed set and auxiliary sets, we can control the expanding directions of multiple sets. Specifically, context features are scored by how well they can tell different sets apart, and the algorithm drives the expanding direction away from ambiguous areas. We demonstrate the effectiveness of SetCoExpan in the below Section.

2.5.2 Probing Knowledge from Pre-trained Language Models

Besides using auxiliary sets to guide the set expansion process, we also explore how to leverage the target class name to enhance SetExpan. Intuitively, knowing the class name is "country", instead of "state" or "city", can help us identify *unambiguous patterns* and eliminate erroneous concepts like "*Europe*" and "*New York*". Moreover, we can acquire such knowledge (i.e., positive and negative class names) by probing a language model automatically without relying on human annotated data.

Motivated by the above intuition, we propose CGExpan, an iterative framework that empowers concept set expansion with class names automatically generated from pre-trained language models (PLMs) [5, 37]. CGExpan consists of three modules: (1) The first, *class name generation* module, constructs and submits class-probing queries (e.g., "[MASK] such as Illinois, Georgia, and Virginia." as shown in Fig. 2.9) to a language model for retrieving a set of candidate class names. (2) The second, *class name ranking* module, builds a concept-probing query for each candidate class name and retrieves a set of concepts. The similarity between this retrieved set and the current concept set serves as a proxy for the class name quality, based on which we rank all candidate class names. (3) The third, *class-guided concept selection* module, scores each concept conditioned on the above selected class names and add top-ranked concepts into the currently expanded set. As better class names may emerge

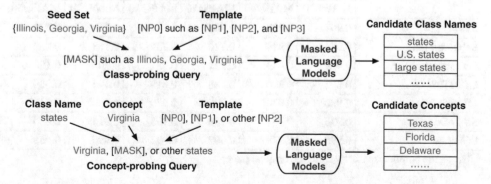

Fig. 2.9 Examples of class-probing and concept-probing queries in CGExpan

in later iterations, we score and rank all concepts (including those already in the expanded set) at each iteration, which helps alleviate the semantic drift issue.

We evaluate the performance of SetCoExpan and CGExpan on Wiki and APR datasets used in previous SetExpan experiments. Besides the previous EgoSet baseline, we further compare SetCoExpan and CGExpan with the following methods:

- SetExpander [17]: This method trains different embeddings based on different types of context features and leverages additional human-annotated sets to build a classifier on top of learned embeddings to predict whether a concept belongs to the set.
- CaSE [38]: This method combines skip-pattern features and embedding features to score and rank concepts once from the corpus. The original paper has three variants and we use the CaSE-W2V variant since it is the best model claimed in the paper.
- MCTS [36]: This method bootstraps the initial seed set by combing the Monte Carlo Tree Search algorithm with a deep similarity network to estimate delayed feedback for pattern evaluation and to score concepts given selected patterns.

Table 2.3 shows the overall performance of different methods. We can see that SetCoExpan and CGExpan in general outperform all the baselines by a large margin. On the Wiki dataset, both SetCoExpan and CGExpan achieve over 24% improvements over SetExpan in MAP@50,. On the APR dataset, CGExpan obtains over 49% improvements over SetExpan in terms of MAP@50. These results verify the effectiveness of parallel expansion of auxiliary sets (in SetCoExpan) and PLM-guided expansion model (in CGExpan).

Table 2.3 Mean average precision on Wiki and APR

Methods	Wiki			APR		
	MAP@10	MAP@20	MAP@50	MAP@10	MAP@20	MAP@50
EgoSet [24]	0.904	0.877	0.745	0.758	0.710	0.570
SetExpan [25]	0.944	0.921	0.720	0.789	0.763	0.639
SetExpander [17]	0.499	0.439	0.321	0.287	0.208	0.120
CaSE [38]	0.897	0.806	0.588	0.619	0.494	0.330
MCTS [36]	0.980[a]	0.930[a]	0.790[a]	0.960[a]	0.900[a]	0.810[a]
SetCoExpan	0.976	0.964	**0.905**	0.933	0.915	0.830
CGExpan	**0.995**	**0.978**	0.902	**0.992**	**0.990**	**0.955**

[a] Means the number is directly from the original paper

2.6   Summary

In this chapter, we present an iterative concept set expansion framework SetExpan with a ranking-based unsupervised ensemble technique for robust concept selection. Our extensive experiments show SetExpan is domain-independent, outperforms many other set expansion methods, and derives high-quality concept sets with minimal human efforts. Furthermore, we introduce a few ways to extend SetExpan by incorporating pre-trained language models and exploiting automatically discovered auxiliary sets.

For future work, it is interesting to (1) extend SetExpan to multiple languages, (2) develop interactive methods to allow users to directly control the expansion process and/or to provide valuable feedbacks after the iterative expansion process ends, and (3) enable the downstream applications of set expansion to provide explicit or implicit feedbacks to guide the concept set expansion model learning.

References

1. Balasubramanyan, R., Dalvi, B.B., Cohen, W.W.: From topic models to semi-supervised learning: Biasing mixed-membership models to exploit topic-indicative features in entity clustering. In: Proceedings of 2013 Joint European Conference on Machine Learning and Knowledge Discovery in Databases (2013)
2. Chen, Z., Cafarella, M., Jagadish, H.: Long-tail vocabulary dictionary extraction from the web. In: Proceedings of the 9th ACM International Conference on Web Search and Data Mining (2016)
3. Chierichetti, F., Kumar, R., Pandey, S., Vassilvitskii, S.: Finding the jaccard median. In: Proceedings of the 21st Annual ACM-SIAM Symposium on Discrete Algorithms (2010)
4. Curran, J.R., Murphy, T., Scholz, B.: Minimising semantic drift with mutual exclusion boot-strapping. In: Proceedings of the 10th Conference of the Pacific Association for Computational Linguistics (2007)
5. Devlin, J., Chang, M.W., Lee, K., Toutanova, K.: BERT: Pre-training of deep bidirectional trans-formers for language understanding. In: Proceedings of the 2019 Conference of the North American Chapter of the Association for Computational Linguistics: Human Language Technologies (2019)
6. Ghahramani, Z., Heller, K.A.: Bayesian sets. In: Proceedings of the 19th Conference on Neural Information Processing Systems (2005)
7. Gupta, S., MacLean, D.L., Heer, J., Manning, C.D.: Research and applications: induced lexico-syntactic patterns improve information extraction from online medical forums. J Amer Med Inform Assoc (2014)
8. Gupta, S., Manning, C.D.: Improved pattern learning for bootstrapped entity extraction. In: Proceedings of the 18th Conference on Computational Natural Language Learning (2014)
9. Gupta, S., Manning, C.D.: Distributed representations of words to guide bootstrapped entity classifiers. In: Proceedings of the 2015 Conference of the North American Chapter of the Association for Computational Linguistics: Human Language Technologies (2015)
10. He, Y., Xin, D.: SEISA: set expansion by iterative similarity aggregation. In: Proceedings of the 20th International Conference on World Wide Web (2011)
11. Huang, J., Xie, Y., Meng, Y., Shen, J., Zhang, Y., Han, J.: Guiding corpus-based set expansion by auxiliary sets generation and co-expansion. In: Proceedings of the 2020 Web Conference (2020)

12. Jindal, P., Roth, D.: Learning from negative examples in set-expansion. In: Proceedings of IEEE 11th International Conference on Data Mining (2011)
13. Lin, D., Wu, X.: Phrase clustering for discriminative learning. In: Proceedings of the 47th Annual Meeting of the Association for Computational Linguistics (2009)
14. Lin, W., Yangarber, R., Grishman, R.: Bootstrapped learning of semantic classes from positive and negative examples. In: Proceedings of ICML-2003 Workshop on The Continuum from Labeled to Unlabeled Data (2003)
15. Ling, X., Weld, D.S.: Fine-grained entity recognition. In: Proceedings of the 2012 AAAI Conference on Artificial Intelligence (2012)
16. Liu, J., Shang, J., Wang, C., Ren, X., Han, J.: Mining quality phrases from massive text corpora. In: Proceedings of the 2015 ACM SIGMOD International Conference on Management of Data (2015)
17. Mamou, J., Pereg, O., Wasserblat, M., Eirew, A., Green, Y., Guskin, S., Izsak, P., Korat, D.: Term set expansion based NLP Architect by Intel AI Lab. In: Proceedings of the 2018 Conference on Empirical Methods in Natural Language Processing (2018)
18. McIntosh, T., Curran, J.R.: Weighted mutual exclusion bootstrapping for domain independent lexicon and template acquisition. In: Proceedings of the Australasian Language Technology Association Workshop 2008 (2008)
19. Mikolov, T., Sutskever, I., Chen, K., Corrado, G.S., Dean, J.: Distributed representations of words and phrases and their compositionality. In: Proceedings of the 27th Conference on Neural Information Processing Systems (2013)
20. Pantel, P., Crestan, E., Borkovsky, A., Popescu, A.M., Vyas, V.: Web-scale distributional similarity and entity set expansion. In: Proceedings of the 2009 Conference on Empirical Methods in Natural Language Processing (2009)
21. Ren, X., El-Kishky, A., Wang, C., Tao, F., Voss, C.R., Han, J.: ClusType: effective entity recognition and typing by relation phrase-based clustering. In: Proceedings of the 24th International Conference on World Wide Web (2015)
22. Ren, X., Lv, Y., Wang, K., Han, J.: Comparative document analysis for large text corpora. In: Proceedings of the 10th ACM International Conference on Web Search and Data Mining (2017)
23. Riloff, E.: Automatically generating extraction patterns from untagged text. In: Proceedings of the 1996 AAAI Conference on Artificial Intelligence (1996)
24. Rong, X., Chen, Z., Mei, Q., Adar, E.: Egoset: exploiting word ego-networks and user-generated ontology for multifaceted set expansion. In: Proceedings of the 9th ACM International Conference on Web Search and Data Mining (2016)
25. Shen, J., Wu, Z., Lei, D., Shang, J., Ren, X., Han, J.: SetExpan: corpus-based set expansion via context feature selection and rank ensemble. In: Proceedings of the 2017 Joint European Conference on Machine Learning and Knowledge Discovery in Databases (2017)
26. Shi, B., Zhang, Z., Sun, L., Han, X.: A probabilistic co-bootstrapping method for entity set expansion. In: Proceedings of the 25th International Conference on Computational Linguistics (2014)
27. Shi, S., Zhang, H., Yuan, X., Wen, J.R.: Corpus-based semantic class mining: distributional vs. pattern-based approaches. In: Proceedings of the 23rd International Conference on Computational Linguistics (2010)
28. Talukdar, P.P., Reisinger, J., Pasca, M., Ravichandran, D., Bhagat, R., Pereira, F.: Weakly-supervised acquisition of labeled class instances using graph random walks. In: Proceedings of the 2008 Conference on Empirical Methods in Natural Language Processing (2008)
29. Tang, J., Qu, M., Mei, Q.: PTE: predictive text embedding through large-scale heterogeneous text networks. In: Proceedings of the 21st ACM SIGKDD International Conference on Knowledge Discovery and Data Mining (2015)

30. Thelen, M., Riloff, E.: A bootstrapping method for learning semantic lexicons using extraction pattern contexts. In: Proceedings of the 2002 Conference on Empirical Methods in Natural Language Processing (2002)
31. Tong, S., Dean, J.: System and methods for automatically creating lists (2008). US Patent 7,350,187
32. Velardi, P., Faralli, S., Navigli, R.: Ontolearn reloaded: a graph-based algorithm for taxonomy induction. In: Computational Linguistics (2013)
33. Wang, C., Chakrabarti, K., He, Y., Ganjam, K., Chen, Z., Bernstein, P.A.: Concept expansion using web tables. In: Proceedings of the 24th International Conference on World Wide Web (2015)
34. Wang, R.C., Cohen, W.W.: Language-independent set expansion of named entities using the web. In: Proceedings of the 7th IEEE International Conference on Data Mining (2007)
35. Wang, Y.Y., Hoffmann, R., Li, X., Szymanski, J.: Semi-supervised learning of semantic classes for query understanding: from the web and for the web. In: Proceedings of the 18th ACM International Conference on Information and Knowledge Management (2009)
36. Yan, L., Han, X., Sun, L., He, B.: Learning to bootstrap for entity set expansion. In: Proceedings of the 2019 Conference on Empirical Methods in Natural Language Processing (2019)
37. Yang, Z., Dai, Z., Yang, Y., Carbonell, J., Salakhutdinov, R., Le, Q.V.: XLNet: generalized autoregressive pretraining for language understanding. In: Proceedings of the 33rd Conference on Neural Information Processing Systems (2019)
38. Yu, P., Huang, Z., Rahimi, R., Allan, J.D.: Corpus-based set expansion with lexical features and distributed representations. In: Proceedings of the 42nd International ACM SIGIR Conference on Research & Development in Information Retrieval (2019)
39. Zhang, Y., Shen, J., Shang, J., Han, J.: Empower entity set expansion via language model probing. In: Proceedings of the 58th Annual Meeting of the Association for Computational Linguistics (2020)

Taxonomy Construction

3

3.1 Overview and Motivations

Concept taxonomy is the backbone of many knowledge-rich applications such as question answering [49], query understanding [14], and personalized recommendation [52]. Most existing concept taxonomies (e.g., MeSH [19], ACM CCS [7], Pinterest Taxonomy [11], etc.) are constructed by human experts or in a crowd-sourcing manner. However, such manual constructions are labor-intensive, time-consuming, unadaptable to changes, and rarely complete. As a result, automated concept taxonomy construction is in great demand.

Existing methods mostly build concept taxonomies based on the "is-A" relation (e.g., a "*panda*" is a "*mammal*") [43, 45, 48] or cluster terms into hierarchically organized topics [9, 36, 44]. However, such hierarchies cannot satisfy many real-world needs due to its (1) *inflexible semantics*: many applications may need hierarchies carrying more flexible semantics such as "*city-state-country*" in a location taxonomy; and (2) *limited applicability*: the "universal" taxonomy so constructed is unlikely to fit diverse and user-specific application tasks. This motivates us to work on the *task-guided* taxonomy construction, which takes a user-provided "seed" taxonomy tree (as task guidance) along with a domain-specific corpus and generates a desired taxonomy automatically. For example, as shown in Fig. 3.1, a user may provide a seed taxonomy containing only two countries and two states along with a large corpus, and our method will output a taxonomy which covers all the countries and states mentioned in the corpus.

In this chapter, we propose HiExpan, a novel framework for task-guided concept taxonomy construction. HiExpan can automatically generates a key term list[1] from the input corpus and iteratively grows the seed taxonomy. Specifically, HiExpan views all children under each taxonomy node forming a coherent set and builds the taxonomy by recursively

---

[1] In this chapter, we use the word "*term*" and "*concept*" interchangeably.

J. Shen and J. Han, *Automated Taxonomy Discovery and Exploration*,
Synthesis Lectures on Data Mining and Knowledge Discovery,
https://doi.org/10.1007/978-3-031-11405-2_3

expanding all these sets using SetExpan. While such an approach is intuitive, there are two major challenges by utilizing SetExpan to generating high-quality taxonomies: (1) modeling global taxonomy information: a concept that appears in multiple expanded sets may need conflict resolution and hierarchy adjustment accordingly, and (2) cold-start with empty initial seed set: as an example, initial seed set {"*Ontario*", "*Quebec*"} will need to be found once we add "*Canada*" at the country level as shown in Fig. 3.1.

HiExpan consists of two novel modules for dealing with the above two challenges. First, whenever we observe a conflict (i.e., the same concept appearing in multiple positions on the taxonomy) during the tree expansion process, we measure a "confidence score" for putting the term in each position and select the most confident position for it. Furthermore, at the end of our hierarchical tree expansion process, we will do a global optimization of the whole tree structure. Second, we incorporate a weakly-supervised relation extraction method to infer parent-child relation information and to find seed children concepts under a specific parent concept. Equipped with these two modules, HiExpan constructs the task-guided taxonomy by iteratively growing the initial seed taxonomy tree. At each iteration, it views all children under a non-leaf taxonomy node as a coherent set and builds the taxonomy by recursively expanding these sets. Whenever a node with no initial children nodes found, it will first conduct seeds hunting. At the end of each iteration, HiExpan detects all the conflicts and resolves them based on their confidence scores.

We summarize our major contributions as follows:

- We introduce a new research problem *task-guided taxonomy construction*, which takes a user-provided seed taxonomy along with a domain-specific corpus as input and aims to output a desired taxonomy that satisfies user-specific application tasks.
- We propose HiExpan, a novel expansion-based framework for task-guided taxonomy construction. HiExpan requires minimum human annotations and generates the taxonomy by growing the seed taxonomy iteratively. Special mechanisms are also taken by HiExpan to leverage global tree structure information.
- We conduct extensive experiments to verify the effectiveness of HiExpan on three real-world datasets from different domains.

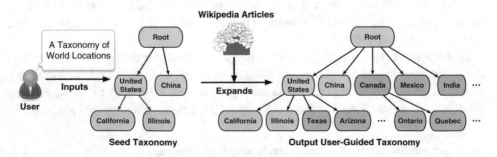

Fig. 3.1 An example of task-guided taxonomy construction

The remaining of this chapter is organized as follows. Section 3.2 discusses the related work. Section 3.3 discuss the task-guided concept taxonomy construction problem and presents our HiExpan framework. We discuss the experiment results in Sect. 3.4 and conclude this chapter with discussions on future research directions in Sect. 3.5.

3.2   Related Work

There are three major lines of previous studies relevant to the current work.

Concept Taxonomy Construction. Most existing approaches to taxonomy construction focus on building hypernym-hyponym taxonomies wherein each parent-child pair expresses the "is-a" relation. Typically, they consist of two key steps: (1) hypernymy relation acquisition (i.e., obtaining hypernym-hyponym pairs), and (2) structured taxonomy induction (i.e., organizing all hypernymy relations into a tree structure).

Methods for hypernymy relation acquisition fall into two classes: pattern-based and distributional. One pioneering pattern-based method is Hearst patterns [13] in which lexical syntactic patterns (e.g., "NP_x such as NP_y") are leveraged to match hypernymy relations. Later studies extend this method by incorporating more linguistic rules [21, 32, 40] or designing generalized patterns such as "*star-pattern*" [28], "*SOL pattern*" [27], and "*meta-pattern*" [15]. These methods could achieve high precision in the result pairs but often suffer low recalls (i.e., many hypernym-hyponym pairs do not match the pre-defined patterns). Along another line, distributional methods predict whether a pair of terms $\langle x, y \rangle$ holds a hypernymy relation based on their distributional representations. Early studies first extract statistical features (e.g., the context words of a term), calculate pairwise term similarity using symmetric metrics (e.g., cosine, Jaccard) [17] or asymmetric metrics (e.g., WeedsPrec [47], SLQS [33]), and predict if $\langle x, y \rangle$ holds a hypernymy relation. More recently, a collections of *supervised* methods [1, 3, 10, 22, 46] are proposed to leverage pre-trained word embeddings and curated training data to directly learn a relation classification/prediction model. However, neither pattern-based nor distributional techniques can be applied to our problem because they are designed exclusively for acquiring hypernym-hyponym pairs, whereas we aim to construct a *task-guided* taxonomy where the parent-child relations are task-specific and subject to user guidance.

For the structured taxonomy induction step, most methods first build a graph where edges represent noisy hypernymy relations, extracted in the former step, and then derive a tree-like taxonomy from this graph. Kozareva and Hovy [16] iteratively retain the longest paths between root and leaf terms and remove other conflicting edges. Velardi et al. [43] use the same longest-path idea to weigh edges and then find the largest-weight taxonomy as a Maximum Spanning Tree. Bansal et al. [2] build a factor graph to model hypernymy relations and regard taxonomy induction as a structured learning problem, which can be inferred with loop belief propagation. Recently, Gupta et al. [12] propose to build the initial graph using hypernym subsequence (instead of single hypernym pair) and model taxonomy induction as a minimum-cost flow problem. Comparing with these methods, our approach leverages the weak supervision in "seed" taxonomy and builds a task-specific taxonomy in which two terms can hold a non-hypernymy relation. Further, our taxonomy construction framework jointly acquires task-specific relations and induces taxonomy structure, instead of performing the two tasks separately.

Topic Hierarchy Construction. There are a number of methods proposed for automatic topic hierarchy construction from text corpora. In pioneer studies, hierarchical topic modeling [4, 9, 39, 44] and bottom-up agglomerative clustering [20, 41] are two most popular frameworks. More recently, after word embedding technique [25] becomes mature, more top-down hierarchical clustering methods [36, 51] are proposed and achieve the new state-of-the-art. All these methods construct a topic hierarchy where each node is represented by a cluster of terms. However, finding a single concept term to summarize the term cluster is proved to be a non-trivial task [8]. In comparison, our HiExpan framework tries to construct a concept taxonomy where each node is naturally represented by a single term.

Weakly-Supervised Relation Extraction. There have been studies on weakly supervised relation extraction, which aims at extracting a set of relation instances containing certain semantic relationships. Our method is related to corpus-level relation extraction that identifies relation instances from the entire text corpora [26, 30, 31, 50]. In the weakly supervised setting, there are generally two approaches for corpus-level relation extraction. The first is pattern-based [15, 27], which usually uses bootstrapping to iteratively extract textual patterns and new relation instances. The second approach [25, 29, 42] tries to learn low-dimensional representations of concepts such that concepts with similar semantic meanings have similar representations. Unfortunately, all these existing methods require a considerable amount of relation instances to train an effective relation classifier, which is infeasible in our setting as we only have a limited number seeds specified by users. Furthermore, these studies do not consider organizing the relation pairs into a taxonomy structure.

3.3   HiExpan: Task-Guided Concept Taxonomy Construction

3.3.1   Problem Formulation

The input for our concept taxonomy construction framework includes two parts: (1) a text corpus $\mathcal{D}$; and (2) a "seed" taxonomy $\mathcal{T}^0$. The "seed" taxonomy $\mathcal{T}^0$, given by a user, is a tree-structured hierarchy and serves as the *task guidance*. Given the corpus $\mathcal{D}$, we aim to expand this seed taxonomy $\mathcal{T}^0$ into a more complete taxonomy $\mathcal{T}$ for the task. Each node $e \in \mathcal{T}$ represents a concept extracted from corpus $\mathcal{D}$ and each edge $\langle e_1, e_2 \rangle$ denotes a pair of concepts that satisfies the task-specific relation. We use $\mathcal{E}$ and $\mathcal{R}$ to denote all the nodes and edges in $\mathcal{T}$ and thus $\mathcal{T} = (\mathcal{E}, \mathcal{R})$.

Figure 3.1 shows an example of our problem. Given a collection of Wikipedia articles (i.e., $\mathcal{D}$) and a "seed" taxonomy containing two countries and two states in the "*U.S.*" (i.e.,

$\mathcal{T}^0 = (\mathcal{E}^0, \mathcal{R}^0))$, we aim to output a taxonomy $\mathcal{T}$ which covers all countries and states mentioned in corpus $\mathcal{D}$ and connects them based on the task-specific relation "*located in*", indicated by $\mathcal{R}^0$.

3.3.2 Framework Overview

In short, HiExpan views all children under each taxonomy node forming a coherent *set*, and builds the taxonomy by recursively expanding all these sets. As shown in Fig. 3.2, two first-level nodes (i.e., "*U.S.*" and "*China*") form a set representing the semantic class "*Country*" and by expanding it, we can obtain all the other countries. Similarly, we can expand the set {"*California*", "*Illinois*"} to find all the other states in the U.S.

Given a corpus, HiExpan first extracts all key terms using a phrase mining tool followed by part-of-speech filtering. As the generated term list contains many irrelevant terms (e.g., people's names are totally irrelevant to a location taxonomy), we use a set expansion technique to carefully select best terms. We refer this process as *width expansion* as it increases the *width* of taxonomy tree. Furthermore, to address the challenge that some nodes do not have an initial child (e.g., the node "*Mexico*" in Fig. 3.2), we find the "seed" children by applying a weakly-supervised relation extraction method, which we refer as *depth expansion*. By iteratively applying these two expansion modules and resolving possible conflicts, our hierarchical tree expansion algorithm will first grow the taxonomy to its full size. Finally, we adjust the taxonomy tree by optimizing its global structure.

3.3.3 Key Term Extraction

We use AutoPhrase, a state-of-the-art phrase mining algorithm [35], to extract all key terms in the given corpus. After that, we apply a Part-of-Speech (POS) tagger to the corpus and obtain the POS tag sequence of each key term occurrence. Then, we retain the key term occurrence whose corresponding POS tag sequence contains a noun POS tag (e.g., "*NN*", "*NNS*", "*NNP*"). Finally, we aggregate the key terms that have at least one remaining occurrence in the corpus into the key term list.

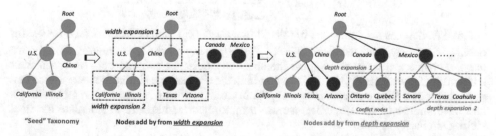

Fig. 3.2 An overview of our hierarchical tree expansion algorithm in HiExpan

3.3.4 Iterative Width and Depth Expansion

3.3.4.1 Width Expansion

idth expansion aims to find the sibling nodes of a given set of children nodes which share the same parent. This naturally forms a set expansion problem and thus we adapt the SetExpan algorithm for addressing it.

Features. We use three types of concept features during the width expansion process:

- *skip-pattern*: Given a target concept e_i in a sentence, one of its skip-pattern features is "w_{-1} _ w_1" where w_{-1} and w_1 are two context words and e_i is replaced with a placeholder. One advantage of skip-pattern feature is that it imposes strong positional constraints. Following [34, 37], we extract up to six skip-patterns of different lengths for one target concept e_i in each sentence.
- *concept embedding*: We use the SkipGram model in word2vec [25] to learn the concept embeddings. For a multi-gram term (e.g., "*Baja California*"), we first use "_" to con-catenate tokens and then learn the embedding of this concept. The advantage of concept embedding feature is that it captures the semantics of each concept.
- *concept type*: We obtain each concept type information by linking it to Probase [48]. The return types serve as the features of that concept. For concepts that are not linkable, they simply do not have this concept's type feature.

Similarity Measures. With above defined concept features, we can compute the sibling similarity of two concepts e_1 and e_2, denoted as $sim_{sib}(e_1, e_2)$. We first assign the weight between each pair of concept and skip-pattern as: $f_{e,sk} = \log(1 + X_{e,sk}) \left[\log |V| - \log(\sum_{e'} X_{e',sk})\right]$, where $X_{e,sk}$ is the raw co-occurrence count between concept e and skip-pattern sk, and $|V|$ is the total number of candidate concepts. Similarly, we can define the association weight between a concept and a type as: $f_{e,ty} = \log(1 + C_{e,ty}) \left[\log |V| - \log(\sum_{e'} C_{e',ty})\right]$, where $C_{e,ty}$ is the confidence score returned by Probase and indicates how confident it believes that concept e has a type ty.

After that, we calculate the similarity of two sibling concepts using skip-pattern features as follows:

$$sim_{sib}^{sk}(e_1, e_2|SK) = \frac{\sum_{sk \in SK} \min(f_{e_1,sk}, f_{e_2,sk})}{\sum_{sk \in SK} \max(f_{e_1,sk}, f_{e_2,sk})}, \tag{3.1}$$

where SK denotes a selected set of "discriminative" skip-pattern features (see below for details). Similarly, we can calculate $sim_{sib}^{tp}(e_1, e_2)$ using all the type features. Finally, we use the cosine similarity to compute the similarity between two concepts based on their embedding features $sim_{sib}^{emb}(e_1, e_2)$. To combine the above three similarities, we notice that a good pair of sibling concepts should appear in similar contexts, share similar embeddings, and have similar types. Therefore, we use a multiplicative measure to calculate the final sibling similarity as follows:

$$sim_{sib}(e_1, e_2|SK) = \sqrt{(1 + sim_{sib}^{sk}(e_1, e_2|SK)) \cdot sim_{sib}^{emb}(e_1, e_2)} \cdot \sqrt{1 + sim_{sib}^{tp}(e_1, e_2)}. \tag{3.2}$$

The Width Expansion process. Given a seed concept set S and a candidate concept list V, a straightforward idea to compute each candidate concept's average similarity with all concepts in the seed set S using all the features. However, this approach can be problematic because (1) the feature space is huge (i.e., there are millions of possible skip-pattern features) and noisy, and (2) the candidate concept list V is also noisy in the sense that many concepts in V are completely irrelevant to S. Therefore, we take a more conservative approach by first selecting a set of quality skip-pattern features and then scoring a concept only if it is associated with at least one quality skip-pattern feature.

Starting with the seed set S, we first score each skip-pattern feature based on its accumulated strength with concepts in S (i.e., $score(sk) = \sum_{e \in S} f_{e,sk}$), and then select top 200 skip-pattern features with maximum scores. After that, we use sampling without replacement method to generate 10 subsets of skip-pattern features $SK_t, t = 1, 2, \ldots, 10$. Each subset SK_t has 120 skip-pattern features. Given an SK_t, we will consider a candidate concept in V only if it has association will at least one skip-pattern feature in SK_t. The score of a considered concept is calculated as follows:

$$score(e|S, SK_t) = \frac{1}{|S|} \sum_{e' \in S} sim_{sib}(e, e'|SK_t).$$
(3.3)

For each SK_t, we can obtain a rank list of candidate concepts L_t based on their scores. We use r_t^i to denote the rank of concept e_i in L_t and if e_i does not appear in L_t, we set $r_t^i = \infty$. Finally, we calculate the mean reciprocal rank (mrr) of each concept e_i and add those concepts with average rank above r into the set S as follows:

$$mrr(e_i) = \frac{1}{10} \sum_{t=1}^{10} \frac{1}{r_t^i}, \quad S = S \cup \{e_i | mrr(e_i) > \frac{1}{r}\}.$$
(3.4)

The key insight of above aggregation mechanism is that an irrelevant concept will not appear frequently in multiple L_t at top positions and thus likely has a low mrr score. The same idea in proved effective in SetExpan and we set $r = 5$ in HiExpan.

3.3.4.2 Depth Expansion

The width expansion algorithm requires an initial seed concept set to start with. This requirement is satisfied for nodes in the initial seed taxonomy $\mathcal{T}^0$ as their children nodes can naturally form such a set. However, for those newly-added nodes in taxonomy tree (e.g., the node "*Canada*" in Fig. 3.2), they do not have any child node and thus we cannot directly apply the width expansion algorithm. To address this problem, we use *depth expansion* algorithm to acquire a target node's initial children by considering the relations between its sibling nodes and its niece/nephew nodes.

Consider the node "*Canada*" in Fig. 3.2 as an example. This node is generated by the previous width expansion algorithm and thus does not have any child node. We aim to find its initial children (i.e., "*Ontario*" and "*Quebec*") by modeling the relation between the siblings of node "*Canada*" (e.g., "*U.S.*") and its niece/nephew node (e.g., "*California*", "*Illinois*"). Similarly, given the target node "*Mexico*", we want to find its initial children such as node "*Sonora*".

Our depth expansion algorithm relies on concept embeddings, which encode the concept semantics in a fix-length dense vector. We use $\mathbf{v}(t)$ to denote the embedding vector of concept t. As shown in [10, 22, 25], the offset of two concepts' embeddings can represent the relationship between them, which leads to the following observation that $\mathbf{v}(\text{"U.S"}) - \mathbf{v}(\text{"California"}) \approx \mathbf{v}(\text{"Canada"}) - \mathbf{v}(\text{"Ontario"})$. Therefore, given a target parent node e_t, a set of reference edges $E = \{\langle e_p, e_c \rangle\}$ where e_p is the parent node of e_c, we calculate the "goodness" of putting node e_x under parent node e_t as follows:

$$
sim_{par}(\langle e_t, e_x \rangle) = cos\left(\mathbf{v}(e_t) - \mathbf{v}(e_x), \frac{1}{|E|} \sum_{\langle e_p, e_c \rangle} \mathbf{v}(e_p) - \mathbf{v}(e_c)\right), \qquad (3.5)
$$

where $cos(\mathbf{v}(x), \mathbf{v}(y))$ denotes the cosine similarity between vector $\mathbf{v}(x)$ and $\mathbf{v}(y)$. Finally, we score each candidate concept e_i based on $sim_{par}(\langle e_t, e_i \rangle)$ and select top-3 concepts with maximum score as the initial children nodes under node e_t.

The concept embedding is learned from REPEL [30], a model for weakly-supervised relation extraction using pattern-enhanced embedding learning. It takes a few seed relation mentions (e.g. "US-Illinois" and "US-California") and outputs concept embeddings as well as reliable relational phrases for target relation type(s). REPEL consists of a pattern module which learns a set of reliable textual patterns, and a distributional module, which learns a relation classifier on concept representations for prediction. As both modules provide extra supervision for each other, the distributional module learns concept embeddings supervised by more reliable patterns from the pattern module. By doing so, the learned concept embeddings carry more useful information than those obtained from other embedding models like word2vec [25], specifically for finding relation tuples of the target relation type(s).

3.3.4.3 Conflict Resolution

We can *iteratively* apply width expansion and depth expansion to grow the taxonomy tree to its full size. As the supervision signal from the user-specified seed taxonomy $\mathcal{T}^0$ is very weak, we need to make sure those nodes introduced in the first several iterations are of high quality and will not mislead the expansion process in later iterations to a wrong direction. In HiExpan, for each task-related concept, we aim to find its single best position on our output task-guided taxonomy $\mathcal{T}$. Therefore, when finding a concept appears in multiple positions during our tree expansion process, we say a "conflict" happens and aim to resolve such conflict by finding the best position that concept should reside in. Given a set of conflicting nodes C which corresponds to different positions of a same concept, we apply the following three rules to select the best node out of this set. First, if any node is in the seed taxonomy $\mathcal{T}^0$, we directly select this node and skip the following two steps. Otherwise, for each pair of nodes in C, we check whether one of them is the ancestor of the other and retain only the ancestor node. After that, we calculate the "confidence score" of each remaining node $e \in C$ as follows:

$$conf(e) = \frac{1}{|sib(e)|} \sum_{e' \in sib(e)} sim_{sib}(e, e'|SK) \cdot sim_{par}(\langle par(e), e \rangle), \qquad (3.6)$$

where $sib(e)$ denotes the set of all sibling nodes of e and $par(e)$ represents its parent node. The skip-pattern feature in SK is selected based on its accumulated strength with concepts in $sib(e)$. The node with highest confidence score will be selected. Finally, for each node in C that is not selected, we will delete the whole subtree rooted by it, cut all the sibling nodes added after it, and put it in its parent node's "children backlist".

In Fig. 3.2, we can see there are two "*Texas*" nodes, one under "*U.S.*" and the other under "*Mexico*". As none of them is from initial "seed" taxonomy and they do not hold an ancestor-descendant relationship, we need to calculate each node's confidence score based on Eq. (3.6). Since "*Texas*" has a stronger relation with other states in U.S., comparing with those in Mexico, we will select the "*Texas*" node under "*U.S.*". Then, for the other node under "*Mexico*", we will delete it and cut "*Coahuila*", a sibling node added after "*Texas*". Finally, we let the node "*Mexico*" to remember that "*Texas*" is not one of its children, which prevents the "*Texas*" node being added back later. Notice that although the "*Coahuila*" node is cut here, it may be added back in a later iteration by our tree expansion algorithm.

3.3.5 Taxonomy Global Optimization

In the above hierarchical tree expansion algorithm, a node will be selected and attached onto the taxonomy based on its "local" similarities with other sibling nodes and its parent node. While modeling only the "local" similarity can simplify the tree expansion process, we find the resulting taxonomy may not be the best from a "global" point of view. For example, when expanding the France regions, we find that "Molise", an Italy region, will be mistakenly added under the "France" node, likely because it shares many similar contexts with some other regions of France. However, when we take a global view of the taxonomy and ask the following question—*which country is Molise located in?*, we can easily put "Molise" under "Italy" as it shares more similarities with those in Italy than in France.

Motivated by the above example, we propose a *taxonomy global optimization module* in HiExpan. The key idea is to adjust each two contiguous levels of the taxonomy tree and to find the best "parent" node at the upper level for each "child" node at the lower level. Our taxonomy global optimization makes the following two hypotheses: (1) concepts that have the same parent are similar to each other and form a coherent set, and (2) each concept is more similar to its correct parent compared with other siblings of its correct parent.

Formally, suppose there are m "parent" nodes at the upper level and n "child" nodes at the lower level, we use $\mathbf{W} \in \mathbb{R}^{n \times n}$ to model the concept-concept sibling similarity and use $\mathbf{Y^c} \in \mathbb{R}^{n \times p}$ to capture the two concepts's parenthood similarity. We let $\mathbf{W_{ij}} = sim_{sib}(e_i, e_j)$ if $i \neq j$, otherwise we set $\mathbf{W_{ii}} = 0$. We set $\mathbf{Y_{ij}^c} = sim_{par}(\langle e_j, e_i \rangle)$. Furthermore, we define another $n \times p$ matrix $\mathbf{Y}^s$ with $\mathbf{Y_{ij}^s} = 1$ if a child node e_i is under parent node e_j and $\mathbf{Y_{ij}^s} = 0$ otherwise. This matrix captures the current parent assignment of each child node. We use $\mathbf{F} \in \mathbb{R}^{n \times p}$ to represent the child nodes' parent assignment we intend to learn. Given a $\mathbf{F}$, we can assign each "child" node e_i to a "parent" node $e_j = \arg \max_j \mathbf{F_{ij}}$. Finally, we propose the following optimization problem to reflect the previous two hypotheses:

$$\min_{\mathbf{F}} \sum_{i,j}^{n} \mathbf{W_{ij}} \left\| \frac{\mathbf{F_i}}{\sqrt{\mathbf{D_{ii}}}} - \frac{\mathbf{F_j}}{\sqrt{\mathbf{D_{jj}}}} \right\|_2^2 + \mu_1 \sum_{i=1}^{n} \left\| \mathbf{F_i} - \frac{\mathbf{Y_i^c}}{\|\mathbf{Y_i^c}\|_1} \right\|_2^2 + \mu_2 \sum_{i=1}^{n} \|\mathbf{F_i} - \mathbf{Y_i^s}\|_2^2, \quad (3.7)$$

where $\mathbf{D_{ii}}$ is the sum of i-th row of $\mathbf{W}$, and μ_1, μ_2 are two nonnegative model hyperparameters. The first term in Eq. (3.7) corresponds to our first hypothesis and models two concepts' sibling similarity. The second term in Eq. (3.7) follows our second hypothesis to model the parenthood similarity. The last term in Eq. (3.7) serves as the smoothness constraints and captures the taxonomy structure information before the global adjustment.

To solve the above optimization problem, we take the derivative of its objective function with respect to $\mathbf{F}$ and can obtain the following closed form solution:

$$\mathbf{F}^* = (\mathbf{I} - \alpha S)^{-1} \cdot (\beta_1 \mathbf{Y^c} + \beta_2 \mathbf{Y^s}), \quad \mathbf{S} = \mathbf{D}^{-1/2} \mathbf{W} \mathbf{D}^{-1/2}, \quad (3.8)$$

where $\alpha_1 = \frac{1}{1+\mu_1+\mu_2}$, $\beta_1 = \frac{\mu_1}{1+\mu_1+\mu_2}$ and $\beta_2 = \frac{\mu_2}{1+\mu_1+\mu_2}$. The calculation procedure is similar to the one in [54].

3.4 Experiments

3.4.1 Datasets

We use two corpora from different domains to evaluate the performance of HiExpan. The first one is **DBLP** which contains about 156 thousand paper abstracts (1.1 million sentences) and a vocabulary of over 17 thousand concepts in the computer science field. The second one is **Wiki** which includes a subset of English Wikipedia pages (1.5 million sentences) used in [18, 37] and a vocabulary of more than 41 thousand concepts.

3.4.2 Compared Methods

To the best of our knowledge, we are the first to study the problem of task-guided taxonomy construction problem, and thus there is no suitable baseline to compare with directly. Therefore, here we evaluate the effectiveness of HiExpan by comparing it with a heuristic set-expansion based method and its own variations as follows:

- HSetExpan is a baseline method which iteratively applies SetExpan algorithm [37] at each level of taxonomy. For each lower level node, this method finds its best parent node to attach according to the children-parent similarity measure defined in Eq. (3.5).
- NoREPEL is a variation of HiExpan without the REPEL [30] module which jointly leverages pattern-based and distributional methods for concept embedding learning. Instead, we use the SkipGram model [25] for learning concept embeddings.
- NoGTO is a variation of HiExpan without the taxonomy global optimization module. It directly outputs the taxonomy generated by hierarchical tree expansion algorithm.
- HiExpan is the full version of our proposed framework, with both REPEL embedding learning module and taxonomy global optimization module enabled.

We use the above methods to generate two taxonomies, one for each corpus. When extracting the key term list using AutoPhrase [35], we treat phrases that occur over 15 times in the corpus to be frequent. The embedding dimension is set to 100 in both REPEL [30] and SkipGram model [25]. The maximum expansion iteration number max_iter is set to 5 for all above methods. Finally, we set the two hyper-parameters used in taxonomy global optimization module as $\mu_1 = 0.1$ and $\mu_2 = 0.01$.

3.4.3 Evaluation Metrics

Evaluating the quality of an entire taxonomy is challenging due to the existence of multiple aspects that should be considered and the difficulty of obtaining gold standard [45]. Following [5, 6], we use **Ancestor-F1** and **Edge-F1** for taxonomy evaluation in this study. **Ancestor-F1** measures correctly predicted ancestral relations. It enumerates all the pairs on the predicted taxonomy and compares these pairs with those in the gold standard taxonomy as follows:

$$P_a = \frac{|\text{is-ancestor}_{\text{pred}} \cap \text{is-ancestor}_{\text{gold}}|}{|\text{is-ancestor}_{\text{pred}}|}, \tag{3.9}$$

$$R_a = \frac{|\text{is-ancestor}_{\text{pred}} \cap \text{is-ancestor}_{\text{gold}}|}{|\text{is-ancestor}_{\text{gold}}|}. \tag{3.10}$$

$$F1_a = \frac{2P_a * R_a}{P_a + R_a} \tag{3.11}$$

Table 3.1 Qualifications of the taxonomies constructed by HSetExpan, NoREPEL, NoGTO, and HiExpan

Methods	Wiki						DBLP					
	P_a	R_a	$F1_a$	P_e	R_e	$F1_e$	P_a	R_a	$F1_a$	P_e	R_e	$F1_e$
HSetExpan	0.740	0.444	0.555	0.759	0.471	0.581	0.743	**0.448**	**0.559**	0.739	0.448	0.558
NoREPEL	0.696	0.596	0.642	0.697	0.576	0.631	0.722	0.384	0.502	0.705	**0.464**	0.560
NoGTO	0.827	0.708	0.763	0.810	0.671	0.734	0.821	0.366	0.506	0.779	0.433	0.556
HiExpan	**0.847**	**0.725**	**0.781**	**0.848**	**0.702**	**0.768**	**0.843**	0.376	0.520	**0.829**	0.460	**0.592**

We denote P_a, R_a, $F1_a$ as the ancestor precision, ancestor recall, and ancestor F1-score, respectively. **Edge-F1** compares edges predicted by different taxonomy construction methods with edges in the gold standard taxonomy. Similarly, we denote edge-based metrics as P_e, R_e, and $F1_e$, respectively.

To construct the gold standard, we extract all the parent-child edges in taxonomies generated by all compared methods. Then we pool all the edges together and ask five people, to judge these pairs independently. We show them seed parent-child pairs as well as the generated parent-child pairs, and ask them to evaluate whether the generated parent-child pairs have the same relation as the given seed parent-child pairs. After collecting these answers from the annotators, we simply use majority voting to label the pairs. We then use these annotated data as the gold standard.

3.4.4   Quantitative Results

Table 3.1 shows both the ancestor-based and edge-based precision/recalls as well as F1-scores of different methods. We can see that HiExpan achieves the best overall performance, and outperforms other methods, especially in terms of the precision. By comparing the performance of HiExpan, NoREPEL, and NoGTO, we can see that both the REPEL module and the taxonomy global optimization algorithm in HiExpan play important roles in improving the quality of the generated taxonomy. Specifically, REPEL learns more discriminative representations by iteratively letting the distributional module and pattern module mutually enhance each other, and the taxonomy global optimization module leverages the global information from the entire taxonomy tree structure. In addition, HiExpan resolves the "conflicts" at the end of each tree expansion iteration by cutting many nodes on a currently expanded taxonomy. This leads HiExpan to generate a smaller tree comparing with the one generated by HSetExpan, given that both methods running the same number of iterations. However, we can see that HiExpan still beats HSetExpan on Wiki dataset, in terms of the recall, which further demonstrates the effectiveness of HiExpan.

3.4.5 Case Studies

In Fig. 3.3, we show the taxonomy trees generated by HiExpan in two domains. First, given a "seed" taxonomy containing two countries and six states/provinces. we can expand it to a full location taxonomy which covers all countries and state/provinces mentioned in the corpus and connects them based the "country-state/province" relation. Similarly, we can expand a seed computer science area taxonomy to automatically discover many other CS subareas. We can also zoom in to look at the taxonomy at a more granular level. Taking the node "natural language processing" as an example, HiExpan successfully finds major subtopics in natural language processing such as "question answering", "text summarization", and "word sense disambiguation" even without any initial seed concepts.

  Table 3.2 shows the effect of taxonomy global optimization module in HiExpan. From the experiment on the Wiki dataset, we observe that the node "London" was originally attached to "Australia", but after applying the taxonomy global optimization module, this node is correctly moved under "England". Similarly, in the DBLP dataset, the term "unsupervised

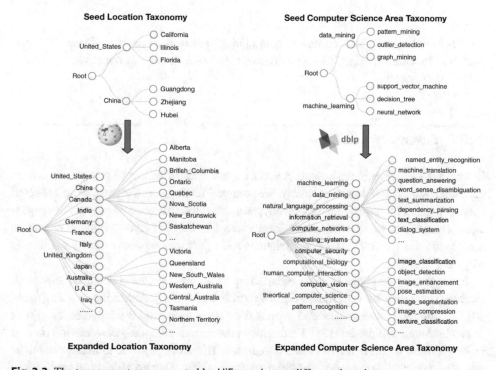

Fig. 3.3 The taxonomy trees generated by HiExpan in two different domains

Table 3.2 NoGTO shows the parent of a concept before applying taxonomy structure optimization. HiExpan shows the parent node of this concept after optimizing the taxonomy structure

Dataset	Concept	NoGTO	HiExpan
Wiki	London	Australia	England
	Chiba	China	Japan
	Molise	Frances	Italy
	New_South_Wales	England	Australia
	Shropshire	Scotland	England
DBLP	Unsupervised_learning	Data_mining	Machine_learning
	Social_network_analysis	Natural_language_processing	Data_mining
	Multi-label_classification	Information_retrieval	Machine_learning
	Pseudo-relevance_feedback	Computational_biology	Information_retrieval
	Function_approximate	Data_analysis	Machine_learning

learning" was initially located under "data mining" but later being moved under the parent node "machine learning". This demonstrates the effectiveness of our taxonomy global optimization module.

3.5   Summary

In this chapter, we explore how to construct a task-guided concept taxonomy based on the initial user-provided seed taxonomy. We propose HiExpan, an effective expansion-based concept taxonomy construction framework that grows the seed taxonomy by recursive expansions. In addition, we discuss how to incorporate a weakly-supervised relation extraction module to infer parent-child concept relations and adjust the taxonomy tree by optimizing its global structure.

In the future, we plan to extend HiExpan by incorporating new concept set expansion methods (including our own studies SetCoExpan and CGExpan), embedding learning methods [23, 24], and more supervision signals from either existing knowledge bases [38] or pre-trained language models [53]. Furthermore, the current taxonomy construction process as well as the evaluation metrics are task-agnostic. Therefore, separate modules need to be designed to apply this taxonomy for different applications. We plan to study an application-guided taxonomy construction method that leverages the performance of a downstream application to guide the upstream taxonomy construction process. Moreover, the current HiExpan takes in user guidances (from the seed taxonomy) only at the initial stage. We plan to extend this framework to allow users provide feedbacks within the whole iterative expansion process. Finally, another interesting direction is to study how the topic taxonomy (where each node contains a set of terms) could be integrated with HiExpan's output single term based taxonomy, which allows for a more flexible and interpretable taxonomy structure.

References

1. Anke, L.E., Camacho-Collados, J., Bovi, C.D., Saggion, H.: Supervised distributional hypernym discovery via domain adaptation. In: Proceedings of the 26th International Conference on Computational Linguistics (2016)
2. Bansal, M., Burkett, D., de Melo, G., Klein, D.: Structured learning for taxonomy induction with belief propagation. In: Proceedings of the 52nd Annual Meeting of the Association for Computational Linguistics (2014)
3. Baroni, M.G., Bernardi, R., Do, N.Q., chieh Shan, C.: Entailment above the word level in distributional semantics. In: Proceedings of the 13th Conference of the European Chapter of the Association for Computational Linguistics (2012)
4. Blei, D., Griffiths, T., Jordan, M.I., Tenenbaum, J.: Hierarchical topic models and the nested chinese restaurant process. In: Proceedings of the 17th Conference on Neural Information Processing Systems (2003)
5. Bordea, G., Buitelaar, P., Faralli, S., Navigli, R.: Semeval-2015 Task 17: taxonomy extraction evaluation (TExEval). In: Proceedings of the 9th International Workshop on Semantic Evaluation (2015)
6. Bordea, G., Lefever, E., Buitelaar, P.: Semeval-2016 Task 13: taxonomy extraction evaluation (TExEval-2). In: Proceedings of the 10th International Workshop on Semantic Evaluation (2016)
7. Cassel, L.N., Palivela, S., Marepalli, S., Padyala, A., Deep, R., Terala, S.: The new ACM CCS and a computing ontology. In: Proceedings of the 13th ACM/IEEE-CS Joint Conference on Digital libraries (2013)
8. Chang, J., Boyd-Graber, J.L., Gerrish, S., Wang, C., Blei, D.: Reading tea leaves: how humans interpret topic models. In: Proceedings of the 23rd Conference on Neural Information Processing Systems (2009)
9. Downey, D., Bhagavatula, C., Yang, Y.: Efficient methods for inferring large sparse topic hierarchies. In: Proceedings of the 53rd Annual Meeting of the Association for Computational Linguistics (2015)
10. Fu, R., Guo, J., Qin, B., Che, W., Wang, H., Liu, T.: Learning semantic hierarchies via word embeddings. In: Proceedings of the 52nd Annual Meeting of the Association for Computational Linguistics (2014)
11. Gonçalves, R.S., Horridge, M., Li, R., Liu, Y., Musen, M.A., Nyulas, C.I., Obamos, E., Shrouty, D., Temple, D.: Use of OWL and semantic web technologies at Pinterest. In: Proceedings of 2019 International Semantic Web Conference (2019)
12. Gupta, A., Lebret, R., Harkous, H., Aberer, K.: Taxonomy induction using hypernym subsequences. In: Proceedings of the 26th ACM International Conference on Information and Knowledge Management (2017)
13. Hearst, M.A.: Automatic acquisition of hyponyms from large text corpora. In: Proceedings of the 15th International Conference on Computational inguistics (1992)
14. Hua, W., Wang, Z., Wang, H., Zheng, K., Zhou, X.: Understand short texts by harvesting and analyzing semantic knowledge. In: IEEE Transactions on Knowledge and Data Engineering (2017)

15. Jiang, M., Shang, J., Cassidy, T., Ren, X., Kaplan, L.M., Hanratty, T.P., Han, J.: MetaPAD: meta pattern discovery from massive text corpora. In: Proceedings of the 23th ACM SIGKDD International Conference on Knowledge Discovery and Data Mining (2017)

16. Kozareva, Z., Hovy, E.H.: A semi-supervised method to learn and construct taxonomies using the web. In: Proceedings of the 2010 Conference on Empirical Methods in Natural Language Processing (2010)

17. Lin, D.: An information-theoretic definition of similarity. In: Proceedings of the 15th International Conference on Machine Learning (1998)

18. Ling, X., Weld, D.S.: Fine-grained entity recognition. In: Proceedings of the 2012 AAAI Conference on Artificial Intelligence (2012)

19. Lipscomb, C.E.: Medical subject headings (MeSH). Bulletin of the Medical Library Association (2000)

20. Liu, X., Song, Y., Liu, S., Wang, H.: Automatic taxonomy construction from keywords. In: Proceedings of the 18th ACM SIGKDD International Conference on Knowledge Discovery and Data Mining (2012)

21. Luu, A.T., jae Kim, J., Ng, S.K.: Taxonomy construction using syntactic contextual evidence. In: Proceedings of the 2014 Conference on Empirical Methods in Natural Language Processing (2014)

22. Luu, A.T., Tay, Y., Hui, S.C., Ng, S.K.: Learning term embeddings for taxonomic relation identification using dynamic weighting neural network. In: Proceedings of the 2016 Conference on Empirical Methods in Natural Language Processing (2016)

23. Meng, Y., Huang, J., Wang, G., Wang, Z., Zhang, C., Zhang, Y., Han, J.: Discriminative topic mining via category-name guided text embedding. Proceedings of the 2020 Web Conference (2020)

24. Meng, Y., Huang, J., Wang, G., Zhang, C., Zhuang, H., Kaplan, L.M., Han, J.: Spherical text embedding. In: Proceedings of the 33rd Conference on Neural Information Processing Systems (2019)

25. Mikolov, T., Sutskever, I., Chen, K., Corrado, G.S., Dean, J.: Distributed representations of words and phrases and their compositionality. In: Proceedings of the 27th Conference on Neural Information Processing Systems (2013)

26. Mintz, M., Bills, S., Snow, R., Jurafsky, D.: Distant supervision for relation extraction without labeled data. In: Proceedings of the 47th Annual Meeting of the Association for Computational Linguistics (2009)

27. Nakashole, N., Weikum, G., Suchanek, F.M.: Patty: A taxonomy of relational patterns with semantic types. In: Proceedings of the 2012 Conference on Empirical Methods in Natural Language Processing (2012)

28. Navigli, R., Velardi, P.: Learning word-class lattices for definition and hypernym extraction. In: Proceedings of the 48th Annual Meeting of the Association for Computational Linguistics (2010)

29. Pennington, J., Socher, R., Manning, C.D.: Glove: Global vectors for word representation. In: Proceedings of the 2014 Conference on Empirical Methods in Natural Language Processing (2014)

30. Qu, M., Ren, X., Zhang, Y., Han, J.: Weakly-supervised relation extraction by pattern-enhanced embedding learning. In: Proceedings of the 27th International Conference on World Wide Web (2018)

31. Riedel, S., Yao, L., McCallum, A., Marlin, B.M.: Relation extraction with matrix factorization and universal schemas. In: Proceedings of the 2013 Conference of the North American Chapter of the Association for Computational Linguistics: Human Language Technologies (2013)

32. Ritter, A., Soderland, S., Etzioni, O.: What is this, anyway: automatic hypernym discovery. In: Papers from the 2009 AAAI Spring Symposium: Learning by Reading and Learning to Read (2009)

33. Roller, S., Erk, K., Boleda, G.: Inclusive yet selective: Supervised distributional hypernymy detection. In: Proceedings of the 25th International Conference on Computational Linguistics (2014)

34. Rong, X., Chen, Z., Mei, Q., Adar, E.: Egoset: exploiting word ego-networks and user-generated ontology for multifaceted set expansion. In: Proceedings of the 9th ACM International Conference on Web Search and Data Mining (2016)

35. Shang, J., Liu, J., Jiang, M., Ren, X., Voss, C.R., Han, J.: Automated phrase mining from massive text corpora. In: IEEE Transactions on Knowledge and Data Engineering (2018)

36. Shang, J., Zhang, X., Liu, L., Li, S., Han, J.: NetTaxo: automated topic taxonomy construction from text-rich network. In: Proceedings of the 2020 Web Conference (2020)

37. Shen, J., Wu, Z., Lei, D., Shang, J., Ren, X., Han, J.: SetExpan: corpus-based set expansion via context feature selection and rank ensemble. In: Proceedings of the 2017 Joint European Conference on Machine Learning and Knowledge Discovery in Databases (2017)

38. Shi, Y., Shen, J., Li, Y., Zhang, N., He, X., Lou, Z., Zhu, Q., Walker, M., Kim, M.H., Han, J.: Discovering hypernymy in text-rich heterogeneous information network by exploiting context granularity. In: Proceedings of the 28th ACM International Conference on Information and Knowledge Management (2019)

39. Shin, S., Moon, I.: Guided HTM: Hierarchical topic model with dirichlet forest priors. IEEE Transactions on Knowledge and Data Engineering (2017)

40. Snow, R., Jurafsky, D., Ng, A.Y.: Learning syntactic patterns for automatic hypernym discovery. In: Proceedings of the 18th Conference on Neural Information Processing Systems (2004)

41. Song, Y., Liu, S., Liu, X., Wang, H.: Automatic taxonomy construction from keywords via scalable bayesian rose trees. In: IEEE Transactions on Knowledge and Data Engineering (2015)

42. Tang, J., Qu, M., Wang, M., Zhang, M., Yan, J., Mei, Q.: LINE: Large-scale information network embedding. In: Proceedings of the 24th International Conference on World Wide Web (2015)

43. Velardi, P., Faralli, S., Navigli, R.: Ontolearn reloaded: a graph-based algorithm for taxonomy induction. In: Computational Linguistics (2013)

44. Wang, C., Danilevsky, M., Desai, N., Zhang, Y., Nguyen, P., Taula, T., Han, J.: A phrase mining framework for recursive construction of a topical hierarchy. In: Proceedings of the 19th ACM SIGKDD International Conference on Knowledge Discovery and Data Mining (2013)

45. Wang, C., He, X., Zhou, A.: A short survey on taxonomy learning from text corpora: issues, resources and recent advances. In: Proceedings of the 2017 Conference on Empirical Methods in Natural Language Processing (2017)

46. Weeds, J., Clarke, D., Reffin, J., Weir, D.J., Keller, B.: Learning to distinguish hypernyms and co-hyponyms. In: Proceedings of the 25th International Conference on Computational Linguistics (2014)

47. Weeds, J., Weir, D.J., McCarthy, D.: Characterising measures of lexical distributional similarity. In: Proceedings of the 20th international conference on Computational Linguistics (2004)

48. Wu, W., Li, H., Wang, H., Zhu, K.Q.: Probase: a probabilistic taxonomy for text understanding. In: Proceedings of the 2012 ACM SIGMOD International Conference on Management of Data (2012)

49. Yang, S., Zou, L., Wang, Z., Yan, J., Wen, J.R.: Efficiently answering technical questions—a knowledge graph approach. In: Proceedings of the 2017 AAAI Conference on Artificial Intelligence (2017)

50. Zeng, D., Liu, K., Chen, Y., Zhao, J.: Distant supervision for relation extraction via piecewise convolutional neural networks. In: Proceedings of the 2015 Conference on Empirical Methods in Natural Language Processing (2015)

51. Zhang, C., Tao, F., Chen, X., Shen, J., Jiang, M., Sadler, B.M., Vanni, M.T., Han, J.: Taxogen: constructing topical concept taxonomy by adaptive term embedding and clustering. In: Proceedings of the 24th ACM SIGKDD International Conference on Knowledge Discovery and Data Mining (2018)
52. Zhang, Y., Ahmed, A., Josifovski, V., Smola, A.J.: Taxonomy discovery for personalized recommendation. In: Proceedings of the 7th ACM International Conference on Web Search and Data Mining (2014)
53. Zhang, Y., Shen, J., Shang, J., Han, J.: Empower entity set expansion via language model probing. In: Proceedings of the 58th Annual Meeting of the Association for Computational Linguistics (2020)
54. Zhou, D., Bousquet, O., Lal, T.N., Weston, J., Schölkopf, B.: Learning with local and global consistency. In: Proceedings of the 17th Conference on Neural Information Processing Systems (2003)

Taxonomy Enrichment

4

4.1 Overview and Motivations

Taxonomies have been fundamental to organizing knowledge for centuries. In today's Web, taxonomies provide valuable knowledge to support many applications such as query understanding [9], content browsing [32], personalized recommendation [10, 40], and web search [17, 31]. For example, many online retailers (e.g., eBay and Amazon) organize products into categories of different granularities, so that customers can easily search and navigate this category taxonomy to find the items they want to purchase. In addition, web search engines (e.g., Google and Bing) leverage a taxonomy to better understand user queries and improve the search quality.

As the web contents and human knowledge are constantly growing, people need to expand an existing taxonomy to include new emerging concepts. Most of previous methods, however, construct a taxonomy entirely *from scratch* and thus when we add new concepts, we have to re-run the entire taxonomy construction process. Although being intuitive, this approach has several limitations. First, many taxonomies have a top-level design provided by domain experts and such design shall be preserved. Second, a newly constructed taxonomy may not be consistent with the old one, which can lead to instabilities of its dependent downstream applications. Finally, as targeting the scenario of building taxonomy from scratch, most previous methods are unsupervised and cannot leverage signals from the existing taxonomy to construct a new one.

In this chapter, we study the *taxonomy expansion* task: given an existing taxonomy and a set of new emerging concepts, we aim to automatically expand the taxonomy to incorporate these new concepts (without changing the existing relations in the given taxonomy). Figure 4.1 shows an example where a taxonomy in computer science domain is expanded to include new subfields (e.g., "*Quantum Computing*") and new techniques (e.g., "*Meta Learning*" and "*UDA*"). Some previous studies [12, 13, 22, 39] attempt this task by using

© The Author(s), under exclusive license to Springer Nature Switzerland AG 2022
J. Shen and J. Han, *Automated Taxonomy Discovery and Exploration*,
Synthesis Lectures on Data Mining and Knowledge Discovery,
https://doi.org/10.1007/978-3-031-11405-2_4

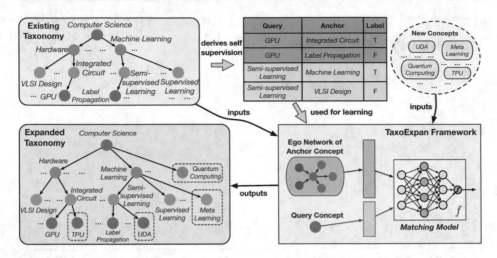

Fig. 4.1 An example of expanding one computer science field-of-studies taxonomy to include new concepts

an additional set of labeled concepts with their true insertion positions in the existing taxonomy. However, such labeled data are usually small and thus forbid us from learning a more powerful model that captures the subsumption semantics in the existing taxonomy.

We propose a novel framework named TaxoExpan to tackle the lack-of-supervision challenge. TaxoExpan formulates a taxonomy as a directed acyclic graph (DAG), automatically generates pseudo-training data from the existing taxonomy, and uses them to learn a matching model for expanding a given taxonomy. Specifically, we view each concept in the existing taxonomy as a *query* and one of its parent concepts as an *anchor*. This gives us a set of positive ⟨query concept, anchor concept⟩ pairs. Then, we generate negative pairs by sampling those concepts that are neither the descendants nor the direct parents of the query concept in the existing taxonomy. In Fig. 4.1, for example, the ⟨*"GPU"*, *"Integrated Circuit"*⟩ is a positive pair and ⟨*"GPU"*, *"Label Propagation"*⟩ is a negative pair. We refer to these training pairs as *self-supervision* data, because they are procedurally generated from the existing taxonomy and no human curation is involved.

To make the best use of above self-supervision data, we develop two novel techniques in TaxoExpan. The first one is a position-enhanced graph neural network (GNN) which encodes the local structure of an anchor concept using its ego network (egonet) in the existing taxonomy. If we view this anchor concept as the "parent" of the query concept, this ego network includes the potential "siblings" and "grand parents" of the query concept. We apply graph neural networks (GNNs) to model this ego network. However, regular GNNs fail to distinguish nodes with different relative positions to the query (i.e., some nodes are grand parents of the query while the others are siblings of the query). To address this limitation, we present a simple but effective enhancement to inject such position information into GNNs using position embedding. We show that such embedding can be easily integrated with existing GNN architectures (e.g., GCN [14] and GAT [29]) and significantly boosts the prediction performance. The second technique is a new noise-robust training scheme based on the InfoNCE loss [19]. Instead of predicting whether each individual ⟨query concept, anchor concept⟩ pair is positive or not, we first group all pairs sharing the same query concept into a single training instance and learn a model to select the positive pair among other negative ones from the group. We show that such training scheme is robust to the label noise and leads to performance gains.

We test the effectiveness of TaxoExpan framework on three real-world taxonomies from different domains. Our results show that TaxoExpan can generate high-quality concept taxonomies in scientific domains and achieves state-of-the-art performance on the WordNet taxonomy expansion challenge [13].

To summarize, our major contributions include:

- We propose a self-supervised framework TaxoExpan that automatically expands existing taxonomies without manually labeled data.
- We develop an effective method for enhancing graph neural network by incorporating hierarchical positional information.
- We design a new training objective that enables the learned model to be robust to label noises in self-supervision data.
- We conduct extensive experiments that verify both the effectiveness and the efficiency of TaxoExpan framework on three real-world taxonomies from different domains.

The rest of this chapter is organized as follows. Section 4.2 discusses the related work. Section 4.3 presents our TaxoExpan framework. Then, Sect. 4.4 reports the experiment results. After that, we discuss extensions of TaxoExpan in Sect. 4.5. Finally, we conclude this chapter in Sect. 4.6.

4.2 Related Work

In many real-world applications, some existing taxonomies may have already been laboriously curated by experts or via crowdsourcing, and are deployed in online systems. Instead of constructing the entire taxonomy from scratch, these applications demand the feature of expanding an existing taxonomy dynamically. There exists some studies on expanding WordNet with named entities from Wikipedia [27] or domain-specific concepts from different corpora [2, 3, 7, 12]. Task 14 of SemEval 2016 challenge [13] is specifically setup to enrich WordNet with concepts from domains like health, sport, and finance. One limitation of these approaches is that they depend on the synset structure unique to WordNet and thus cannot be easily generalized to other taxonomies.

To address the above limitation, more recent works try to develop methodologies for expanding a generic taxonomy. Wang et al. [30] design a hierarchical Dirichlet model to extend the category taxonomy in search engines using query logs. Plachouras et al. [21] learn paraphrase models on external PPDB datasets and apply learned models to directly find paraphrases of concepts in the existing taxonomy. Vedula et al. [28] combine multiple features, some of which are retrieved from an external `Bing` Search API, into a ranking model to score candidate positions in terms of their matching scores with the query concept. Aly et al. [1] first learn term embeddings in a hyperbolic space and then attach each new concept to its most similar node in the existing taxonomy based on the hyperbolic embeddings. Comparing with these methods, our proposed framework (details in the next section) has two advantages. First, it requires no additional resource and makes full use of the existing taxonomy as the self supervision, which leads to a border application scope. Second, it explicitly models the local structure around each candidate position, which boosts the quality of expanded taxonomy.

Our work is also related to Graph Neural Network (GNN) which is a generic method of learning on graph-structure data. Many GNN architectures have been proposed to either learn individual node embeddings [6, 8, 14, 29] for the node classification and the link prediction tasks or learn an entire graph representation [15, 34, 41] for the graph classification task. In this work, we tackle the taxonomy expansion task with a fundamentally different formulation from previous tasks. We leverage some existing GNN architectures and enrich them with additional relative position information. Recently, You et al. [36] propose a method to add position information into GNN. Our methods are different from You et al.. They model the *absolute* position of a node in a full graph without any particular reference points; while our technique captures the *relative* position of a node with respect to the query node. Finally, some work on graph generation [11, 16, 35] involves a module to add a new node into a partially generated graph, which shares the similar goal as our model. However, such graph generation model typically requires fully labeled training data to learn from. To the best of our knowledge, this is the first study on how to expand an existing directed acyclic graph (as we model a taxonomy as a DAG) using self-supervised learning.

4.3 TaxoExpan: Self-supervised Taxonomy Expansion

4.3.1 Problem Formulation

Taxonomy. A taxonomy $\mathcal{T} = (\mathcal{N}, \mathcal{E})$ is a directed acyclic graph where each node $n \in \mathcal{N}$ represents a concept (i.e., a word or a phrase) and each directed edge $\langle n_p, n_c \rangle \in \mathcal{E}$ indicates a relation expressing that concept n_p is the most specific concept that is more general than concept n_c. In other words, we refer to n_p as the "*parent*" of n_c and n_c as the "*child*" of n_p.

Problem Definition. The input of the *taxonomy expansion task* includes two parts: (1) an existing taxonomy $\mathcal{T}^0 = (\mathcal{N}^0, \mathcal{E}^0)$, and (2) a set of new concepts $\mathcal{C}$. This new concept set can be either manually specified by users or automatically extracted from text corpora. Our goal is to expand the existing taxonomy $\mathcal{T}^0$ into a larger taxonomy $\mathcal{T} = (\mathcal{N}^0 \cup C, \mathcal{E}^0 \cup \mathcal{R})$, where $\mathcal{R}$ is a set of newly discovered relations each including one new concept $c \in C$.

Figure 4.1 shows an example of our problem. Given a field-of-study taxonomy $\mathcal{T}^0$ in the computer science domain and a set of new concepts $C = \{$ *"UDA"*, *"Meta Learning"*, $\ldots\}$, we find each new concept's best position in $\mathcal{T}^0$ (e.g., "UDA" under "Semi-supervised Learning" as well as "GPU" under "Integrated Circuit") and expand $\mathcal{T}^0$ to include those new concepts.

Simplified Problem. A simplified version of the above problem is that we assume the input set of new concepts contains only one element (i.e., $|C| = 1$), and we aim to find one single parent node of this new concept (i.e., $|\mathcal{R}| = 1$). We discuss the connection between these two problem settings later in this chapter.

Discussion. In this work, we follow previous studies [1, 13, 28] and assume each concept in $\mathcal{N}^0 \cup C$ has an initial embedding vector learned from this concept's surface name, or if available, its definition sentences [22] and associated web pages [30]. We also note that our problem formulation assumes those relations in the existing taxonomy are not modified. We acknowledge that such modification is necessary in some cases, but it is much less frequent and requires high cautiousness from human curators. Therefore, we leave it out of the scope of automation in this study.

In the below section, we first introduce our taxonomy model and expansion goal. Then, we elaborate how to represent a query concept and an insertion position (i.e., an anchor concept), based on which we present our query-concept matching model. Finally, we discuss how to generate self-supervision data from the existing taxonomy and use them to train the TaxoExpan framework.

4.3.2 Taxonomy Modeling and Expansion Goal

A taxonomy $\mathcal{T}$ describes a hierarchical organization of concepts. These concepts form the node set $\mathcal{N}$ in $\mathcal{T}$. Mathematically, we model each node $n \in \mathcal{N}$ as a categorical random variable and the entire taxonomy $\mathcal{T}$ as a Bayesian network. We define the probability of a taxonomy $\mathcal{T}$ as the joint probability of node set $\mathcal{N}$ which can be further factorized into a set of conditional probabilities as follows:

$$\mathbf{P}(\mathcal{T}|\Theta) = \mathbf{P}(\mathcal{N}|\mathcal{T}, \Theta) = \prod_{i=1}^{|\mathcal{N}|} \mathbf{P}(n_i|parent_{\mathcal{T}}(n_i), \Theta), \quad (4.1)$$

where Θ is the set of model parameters and $parent_{\mathcal{T}}(n_i)$ is the set of n_i's parent node(s) in taxonomy $\mathcal{T}$. Given learned model parameters Θ, an existing taxonomy $\mathcal{T}^0 = (\mathcal{N}^0, \mathcal{E}^0)$, and a set of new concepts C, we can ideally find the best taxonomy $\mathcal{T}^*$ by solving the following optimization problem:

$$\mathcal{T}^* = \arg\max_{\mathcal{T}} \mathbf{P}(\mathcal{T}|\Theta) = \arg\max_{\mathcal{T}} \sum_{i=1}^{|\mathcal{N}^0 \cup C|} \log \mathbf{P}(n_i|parent_{\mathcal{T}}(n_i), \Theta). \tag{4.2}$$

This naïve approach has two limitations. First, the search space of all possible taxonomies over the concept set $|\mathcal{N}^0 \cup C|$ is prohibitively large. Second, we cannot guarantee the structure of existing taxonomy $\mathcal{T}^0$ remains unchanged, which can be undesirable from the application point of view. We address the above limitations by restricting the search space of our output taxonomy to be the exact expansion of the existing taxonomy $\mathcal{T}^0$. Specifically, we keep the parents of each existing taxonomy node $n \in \mathcal{N}^0$ unchanged and only try to find a *single* parent node of each new concept in C. As a result, we divide the above computationally intractable problem into the following set of $|C|$ tractable optimization problems:

$$a_i^* = \arg\max_{a_i \in \mathcal{N}^0} \log \mathbf{P}(n_i|a_i, \Theta), \quad \forall i \in \{1, 2, \ldots, |C|\}, \tag{4.3}$$

where a_i is the parent of a new concept $n_i \in C$ and we refer to it as the "*anchor concept*".

Discussion. The above equation defines $|C|$ *independent* optimization problems and each problem aims to find one single parent of a new concept n_i. Therefore, we essentially reduce the more generic taxonomy expansion problem into $|C|$ independent simplified problems and tackle it by inserting new concepts *one-by-one* into the existing taxonomy. As a result of the above reduction, possible interactions among new concepts are ignored and we leave it to the future work. In the following sections, we continue to answer two keys questions: (1) how to model the conditional probability $\mathbf{P}(n_i|a_i, \Theta)$, and (2) how to learn model parameters Θ.

4.3.3   Query-Anchor Matching Model

We model the matching score between a query concept n_i and an anchor concept a_i by projecting them into a vector space and calculating matching scores using their vectorized representations. We show the entire model architecture of TaxoExpan in Fig. 4.2.

4.3.3.1 Query-Anchor Matching: Query Concept Representation
In this study, we assume each query concept has an *initial feature vector* learned based on some text associated with this concept. Such text can be as simple as the concept surface name, or in some prior studies [13, 30], the definition sentences and clicked web pages about the concept. We represent each query concept n_i using its initial feature vector denoted as $\mathbf{n}_i$. We will discuss how to obtain such initial feature vectors using embedding learning methods in the experiment section.

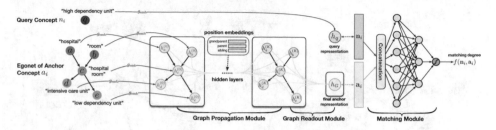

Fig. 4.2 Overview of TaxoExpan framework. g_{emb} is an embedding model that provides query concept's initial feature vector h_q and the initial feature vector of each node in the egonet. The graph propagation module transforms initial feature vectors into better node representations based on which the graph readout module outputs the egonet embedding as the final anchor representation. Finally, a matching module inputs both query and anchor representations and outputs their matching score

4.3.3.2 Query-Anchor Matching: Anchor Concept Representation

Each anchor concept corresponds to one node in the existing taxonomy $\mathcal{T}^0$ that could be the "*parent*" of a query concept. One naïve way to represent an anchor concept is to directly use its initial feature vector. A key limitation of this approach is that it captures only the "*parent*" node information and loses other surrounding nodes' signals, which could be crucial for determining whether the query concept should be put in this position. We illustrate this limitation below.

Suppose we are given a query concept "*high dependency unit*" to predict whether it should be under the "*hospital room*" node (i.e., an anchor concept) in an existing taxonomy. As these two concepts have dissimilar embeddings based on their surface names, we may believe this query concept shouldn't be placed underneath this anchor concept. However, if we know that this anchor concept has two children nodes, i.e., "*intensive care unit*" and "*low dependency unit*", that are closely related to the query concept, we are more likely to put the query concept under this anchor concept, correctly.

The above example demonstrates the importance of capturing local structure information in the anchor concept representation. Thus, we model the anchor concept using its ego network. Specifically, we consider the anchor concept to be the "*parent*" node of a query concept. The ego network of the anchor concept consists of the "*sibling*" nodes and "*grand parent*" nodes of the query concept, as shown in Fig. 4.3. We represent the anchor concept based on its ego network using a graph neural network.

Graph Neural Network Architectures. Given an anchor concept a_i with its corresponding ego network G_{a_i} and its initial representation a_i, we use a graph neural network (GNN) to generate its final representation $\mathbf{a}_i$. This GNN contains two components: (1) a *graph propagation* module that transforms and propagates node features over the graph structure to compute individual node embeddings in G_{a_i}, and (2) a *graph readout* module that combines node embeddings into a vector representing the full ego network G_{a_i}. The final graph embedding encodes all local structure information centered around the anchor concept and we use it as the final anchor representation $\mathbf{a}_i$.

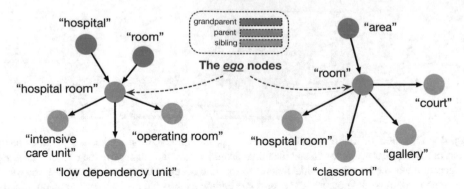

Fig. 4.3 Two egonets correspond to two anchor concepts

A graph propagation module uses a neighborhood aggregation strategy to iteratively update the representation of a node u by aggregating representations of its neighbors $N(u)$ and itself. We denote $N(u) \cup \{u\}$ as $\widetilde{N(u)}$. After K iterations, a node's representation captures the structural information within its K-hop neighborhood. Formally, we define a GNN with K-layers as follows:

$$h_u^{(k)} = \mathrm{AGG}^{(k)}\left(\{h_v^{(k-1)} | v \in \widetilde{N(u)}\}\right), \quad k \in \{1, \dots, K\}, \tag{4.4}$$

where $h_u^{(k)}$ is node u's feature in the kth layer; $h_u^{(0)}$ is node u's initial feature vector, and $\mathrm{AGG}^{(k)}$ is an aggregation function in the kth layer. We instantiate $\mathrm{AGG}^{(k)}$ using two popular architectures: Graph Convolutional Network (GCN) [14] and Graph Attention Network (GAT) [29]. GCN defines the AGG function as follows:

$$\mathrm{AGG}^{(k)}\left(\{h_v^{(k-1)} | v \in \widetilde{N(u)}\}\right) = \rho\left(\sum_{v \in \widetilde{N(u)}} \alpha_{uv}^{(k-1)} \mathbf{W}^{(k-1)} h_v^{(k-1)}\right), \tag{4.5}$$

where $\alpha_{uv}^{(k-1)} = 1/\sqrt{|\widetilde{N(u)}||\widetilde{N(v)}|}$ is a normalization constant (same for all layers); ρ is a non-linear function (e.g., ReLU), and $\mathbf{W}^{(k-1)}$ is the learnable weight matrix. If we interpret $\alpha_{uv}^{(k-1)}$ as the *importance* of node v's feature to node u, GCN calculates it using only the graph structure without leveraging the node features. GAT addresses this limitation by defining $\alpha_{uv}^{(k-1)}$ as follows:

$$\alpha_{uv}^{(k-1)} = \frac{\exp\left(\gamma\left(\mathbf{z}^{(k-1)}[\mathbf{W}^{(k-1)}h_u^{(k-1)} \| \mathbf{W}^{(k-1)}h_v^{(k-1)}]\right)\right)}{\sum_{v' \in \widetilde{N(u)}} \exp\left(\gamma\left(\mathbf{z}^{(k-1)}[\mathbf{W}^{(k-1)}h_u^{(k-1)} \| \mathbf{W}^{(k-1)}h_{v'}^{(k-1)}]\right)\right)}, \tag{4.6}$$

where both $\mathbf{z}^{(k-1)}$ and $\mathbf{W}^{(k-1)}$ are learnable parameters; $\gamma(\cdot)$ is another non-linear function (e.g., LeakyReLU), and "$\|$" represents the concatenation operation. Plugging the above $\alpha_{uv}^{(k-1)}$ into Eq. (4.5) we obtain the aggregation function in a *single-head* GAT. Finally, We execute M independent transformations of Eq. (4.5) and concatenate their output features to compose the final output embedding of node u. This defines the aggregation function in a *multi-head* GAT (with M heads) as follows:

$$\text{AGG}^{(k)}\left(\{h_v^{(k-1)}|v \in \widetilde{N(u)}\}\right) = \overset{M}{\underset{m=1}{\|}} \rho\left(\sum_{v \in \widetilde{N(u)}} \alpha_{uv}^{(k-1)} \mathbf{W}_m^{(k-1)} h_v^{(k-1)}\right), \qquad (4.7)$$

where $\mathbf{W}_m^{(k-1)}$ is the mth weight matrix in the mth attention head.

After obtaining each node's final representation $h_u^{(K)}$, we generate the ego network's representation h_G using a graph readout module as follows:

$$h_G = \text{READOUT}(\{h_u^{(K)}|u \in G\}), \qquad (4.8)$$

where READOUT is a permutation invariant function [38] such as element-wise average, maximum, or summation.

Position-enhanced Graph Neural Networks. One key limitation of the above GNN model is that they fail to capture each node's position information relative to the query concept. For example, in Fig. 4.3, the "*hospital room*" node in the left ego network is the anchor node itself while in the right ego network it is the child of the anchor node. Such position information will influence how node feature propagates within the ego network and how the final graph embedding is aggregated.

An important innovation in TaxoExpan is the design of position-enhanced graph neural networks. The key idea is to learn a set of "position embeddings" and enrich each node feature with its corresponding position embedding. We denote node u's position as p_u and its position embedding at kth layer as $\mathbf{p}_u^{(k)}$. We replace each node feature $h_u^{(k-1)}$ with its position-enhanced version $h_u^{(k-1)}\|\mathbf{p}_u^{(k-1)}$ in Eqs. (4.5)–(4.7) and adjust the dimensionality of $\mathbf{W}^{(k-1)}$ accordingly. Such position embeddings help us to learn better node representations from two aspects. First, we can capture more neighborhood information. Take $\mathbf{W}^{(k-1)}h_v^{(k-1)}$ in the right hand side of Eq. (4.5) as an example, we enhance it to the following:

$$\left[\mathbf{W}^{(k-1)}\|\mathbf{O}^{(k-1)}\right]\left[h_v^{(k-1)}\|\mathbf{p}_v^{(k-1)}\right] = \mathbf{W}^{(k-1)}h_v^{(k-1)} + \mathbf{O}^{(k-1)}\mathbf{p}_v^{(k-1)}, \qquad (4.9)$$

where $\mathbf{O}^{(k-1)}$ is another weight matrix used to transform position embeddings. The above equation shows that a node's new representation is jointly determined by its neighborhoods' contents (i.e., $h_v^{(k-1)}$) and relative positions in the ego network (i.e., $\mathbf{p}_v^{(k-1)}$). Second, for GAT architecture, we can better model neighbor importance as the term $\alpha_{uv}^{(k-1)}$ in Eq. (4.5) currently depends on both $\mathbf{p}_u^{(k-1)}$ and $\mathbf{p}_v^{(k-1)}$.

Furthermore, we propose two schemes to inject position information in the graph readout module. The first one, called weighted mean readout (WMR), is defined as follows:

$$\text{READOUT}(\{h_u^{(K)}|u \in G\}) = \sum_{u \in G} \frac{\log(1 + \exp(\alpha_{p_u}))}{\sum_{u' \in G} \log(1 + \exp(\alpha_{p'_u}))} h_u^{(K)}, \tag{4.10}$$

where α_{p_u} is the parameter indicating the importance of position p_u. The second scheme is called concatenation readout (CR) which combines the average embeddings of nodes with the same position as follows:

$$\text{READOUT}(\{h_u^{(K)}|u \in G\}) = \Big\|_{p \in \mathcal{P}} \frac{\mathcal{I}(p_u = p)h_u^{(K)}}{\sum_{u' \in G} \mathcal{I}(p_{u'} = p)}, \tag{4.11}$$

where $\mathcal{P}$ is the set of all positions we are modeling and $\mathcal{I}(\cdot)$ is an indicator function which returns 1 if its internal statement is true and returns 0 otherwise.

4.3.3.3 Query-Anchor Matching: Matching Model

Based on the learned query concept representation $\mathbf{n}_i \in \mathbb{R}^{D_1}$ and anchor concept representation $\mathbf{a}_i \in \mathbb{R}^{D_2}$, we calculate their match score using a matching module $f(\cdot) : \mathbb{R}^{D_2} \times \mathbb{R}^{D_1} \to \mathbb{R}$. We study two architectures. The first one is a multi-layer perceptron with one hidden layer, defined as follows:

$$f^{\text{MLP}}(\mathbf{a}_i, \mathbf{n}_i) = \sigma\left(\mathbf{W}_2 \gamma\left(\mathbf{W}_1(\mathbf{a}_i \| \mathbf{n}_i) + \mathbf{B}_1\right) + \mathbf{B}_2\right), \tag{4.12}$$

where $\{\mathbf{W}_1, \mathbf{B}_1, \mathbf{W}_2, \mathbf{B}_2\}$ are parameters; $\sigma(\cdot)$ is the sigmoid function, and $\gamma(\cdot)$ is the LeakyReLU activation function. The second architecture is a log-bilinear model defined as follows:

$$f^{\text{LBM}}(\mathbf{a}_i, \mathbf{n}_i) = \exp\left(\mathbf{a}_i^T \mathbf{W} \mathbf{n}_i\right), \tag{4.13}$$

where $\mathbf{W}$ is a learnable interaction matrix. We choose these MLP and LBM as they are representative architectures in linear and bilinear interaction models, respectively.

4.3.4   Model Learning and Inference

The above sections discuss how to model query-anchor matching using a parameterized function $f(\cdot|\Theta)$. Here, we introduce how to learn those parameters using self-supervision from the existing taxonomy and establish the connection between the matching score with the conditional probability $\mathbf{P}(n_i|a_i)$. Finally, we discuss how to conduct model inference.

Self-supervision Generation. Figure 4.4 shows the generation process of self supervision data. Given one edge $\langle n_p, n_c \rangle$ in the existing taxonomy $\mathcal{T}^0 = (\mathcal{N}^0, \mathcal{E}^0)$, we first construct a positive $\langle anchor, query \rangle$ pair by using child node n_c as the "*query*" and parent node n_p as the

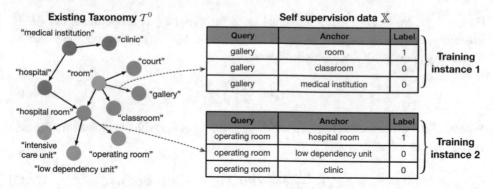

Fig. 4.4 Self-supervision generation in TaxoExpan framework

"*anchor*". Then, we construct N negative pairs by fixing the query node n_c and randomly selecting N nodes $\{n_r^l|_{l=1}^N\} \subset \mathcal{N}^0$ that are neither parents nor descendants of n_c. These $N + 1$ pairs (one positive and N negatives) collectively consist of one training instance $\mathbf{X} = \{\langle n_p, n_c \rangle, \langle n_r^1, n_c \rangle, \ldots, \langle n_r^N, n_c \rangle\}$. By repeating the above process for each edge in $\mathcal{T}^0$, we obtain the full self-supervision dataset $\mathbb{X} = \{\mathbf{X}_1, \ldots, \mathbf{X}_{|\mathcal{E}^0|}\}$. Notice that a node with C parents in $\mathcal{T}^0$ will derive C training instances in $\mathbb{X}$.

Model Training. We learn our model on $\mathbb{X}$ using the InfoNCE loss [19] as follows:

$$\mathcal{L}(\Theta) = -\frac{1}{|\mathbb{X}|} \sum_{\mathbf{X}_i \in \mathbb{X}} \left[\log \frac{f(n_p, n_c)}{\sum_{\langle n_j, n_c \rangle \in \mathbf{X}_i} f(n_j, n_c)} \right], \tag{4.14}$$

where the subscript $j \in [1, 2, \ldots, N + 1]$. If $j = 1$, $\langle n_j, n_c \rangle$ is a positive pair, otherwise, $\langle n_j, n_c \rangle$ is a negative pair. In other words, $\mathbf{X}$ contains query n_c's one positive anchor (i.e., its true parent n_p) sampled from the true distribution $\mathbf{P}(a_i|n_c)$ and N negative anchors $\{n_r^l|_{l=1}^N\}$ sampled from a uniform distribution $\mathbf{P}(a_i)$. If we merge these $N + 1$ anchors into a small set and consider the task of selecting true anchor n_p's position j^* in $[1, 2, \ldots, N + 1]$, we can view Eq. (4.14) as the cross entropy of position distribution $\hat{\mathbf{P}}$ from model prediction relative to the true distribution $\mathbf{P}^*$. Specifically, the model predicted position distribution $\hat{\mathbf{P}}_j = \frac{f(a_j, n_c)}{\sum_{k=1}^{N+1} f(a_k, n_c)}$ where one of $\{a_k|_{k=1}^{N+1}\}$ is the true anchor and all the others are negative anchors. Meanwhile, in the true position distribution:

$$\mathbf{P}_j^* = \frac{\mathbf{P}(a_j|n_c) \prod_{l \neq j} \mathbf{P}(a_l)}{\sum_{k=1}^{N+1} \left(\mathbf{P}(a_k|n_c) \prod_{l \neq k} \mathbf{P}(a_l) \right)} = \frac{\frac{\mathbf{P}(a_j|n_c)}{\mathbf{P}(a_j)}}{\sum_{k=1}^{N+1} \frac{\mathbf{P}(a_k|n_c)}{\mathbf{P}(a_k)}}. \tag{4.15}$$

From above, we can see that the optimal value for $f(a_j, n_c)$ is proportional to $\frac{\mathbf{P}(a_j|n_c)}{\mathbf{P}(a_j)}$. Therefore, optimizing the loss in Eq. (4.14) results in $f(a_i, n_i)$ estimating the following probability density (up to a multiplicative constant):

$$f(a_i, n_i) \propto \frac{\mathbf{P}(a_i|n_i)}{\mathbf{P}(a_i)}. \tag{4.16}$$

We establish the connection between matching score $f(a_i, n_i)$ with the probability $\mathbf{P}(n_i|a_i)$ in Eq. (4.3) as follows:

$$\mathbf{P}(n_i|a_i) = \frac{\mathbf{P}(a_i|n_i)}{\mathbf{P}(a_i)} \cdot \mathbf{P}(n_i) \propto f(a_i, n_i) \cdot \mathbf{P}(n_i). \tag{4.17}$$

We elaborate the implication of the above equation below and summarize our self-learning procedure in Algorithm 4.1.

Model Inference. At the inference stage, we are given a new query concept n_i and apply the learned model $f(\cdot|\Theta)$ to predict its parent node in the existing taxonomy $\mathcal{T}^0$. Mathematically, we aim to find the anchor position a_i that maximizes $\mathbf{P}(n_i|a_i)$, which is equivalent to maximizing $f(a_i, n_i)$ because of Eq. (4.17) and the fact that $P(n_i)$ is the same across all positions. Therefore, we rank all candidate positions a_i based on their matching scores with n_i and select the top ranked one as the predicted parent node of this query concept. Although we currently select only the top one as query's single parent, we can also choose top-k ones as query's parents, if needed.

Summary. Given an existing taxonomy and a set of new concepts, our TaxoExpan framework first generates a set of self-supervision data and learns its internal model parameters using Algorithm 4.1. For each new concept, we run the inference procedure and find its best parent node in the existing taxonomy. Finally, we place these new concepts underneath their predicted parents one at a time, and output the expanded taxonomy.

Computational Complexity Analysis. At the training stage, our model uses $|\mathcal{E}^{(0)}|$ training instances every epoch and thus scales linearly to the number of edges in the existing taxonomy. At the inference stage, for each query concept, we calculate $|\mathcal{N}^{(0)}|$ matching scores, one for every existing node in $\mathcal{T}^0$. Although such $O(|\mathcal{N}^{(0)}|)$ cost per query is expensive, we can significantly reduce it using two strategies. First, most computation efforts of TaxoExpan are matrix multiplications and thus we use GPU for acceleration. Second, as the graph propagation and graph readout modules are query-independent (c.f. Fig. 4.4), we pre-compute and cache all anchor representations. When a set of queries are given, we only run the matching module. In practice, it takes <30 s to calculate all matching scores between 2450 queries with over 24,000 anchor positions on a single K80 GPU.

Algorithm 4.1: Self-supervised learning of TaxoExpan

Input: A taxonomy $\mathcal{T}^0$; negative size N, batch size B; model $f(\cdot|\Theta)$.
Output: Learned model parameters Θ.
1 Randomly initialize Θ;
2 **while** $\mathcal{L}(\Theta)$ *in Eq. (4.14) not converge* **do**
3   Enumerate edges in $\mathcal{T}^0$ and sample B edges without replacement;
4      $\mathbb{X} = \{\}$ # current batch of training instances;
5   **for** *each sampled edge* $\langle n_p, n_c \rangle$ **do**
6   Generate N negative pairs $\{\langle n_r^l, n_c \rangle |_{l=1}^N\}$;
7          $\mathbb{X} \leftarrow \mathbb{X} \cup \{\langle n_p, n_c \rangle, \langle n_r^1, n_c \rangle, \ldots, \langle n_r^N, n_c \rangle\}$;
8   Update Θ based on $\mathbb{X}$.
9 Return Θ;

4.4   Experiments

We conduct two sets of experiments to verify the effectiveness of TaxoExpan framework. Section 4.4.1 presents our results on the Field-of-Study (FoS) Taxonomy[1] in Microsoft Academia Graph (MAG) [25]. Section 4.4.2 presents the results on the SemEval 2016 Task 14 benchmark dataset[2] [13]. Table 4.1 lists the dataset statistics.

4.4.1   Experiments on MAG Dataset

4.4.1.1 Experiment Settings on MAG Datasets

Datasets. We evaluate TaxoExpan on the public Field-of-Study (FoS) Taxonomy in Microsoft Academic Graph (MAG). This FoS taxonomy contains over 660 thousand scientific concepts and more than 700 thousand taxonomic relations. Although being constructed semi-automatically, this taxonomy is of high quality, as shown in the previous study [23]. Thus we treat each concept's original parent nodes as its correct anchor positions. We remove all

Table 4.1 Dataset statistics. $|\mathcal{N}|$ and $|\mathcal{E}|$ are the number of nodes and edges in the existing taxonomy. $|\mathcal{D}|$ indicates the taxonomy depth and $|C|$ is the number of new concepts

Dataset	$|\mathcal{N}|$	$|\mathcal{E}|$	$|\mathcal{D}|$	$|C|$
MAG-CS	24,754	42,329	6	2450
MAG-Full	355,808	638,674	6	37,804
SemEval	95,882	89,089	20	600

---

[1] https://docs.microsoft.com/en-us/academic-services/graph/reference-data-schema.
[2] http://alt.qcri.org/semeval2016/task14/.

concepts that have no relation in the original FoS taxonomy and then randomly mask 20% of leaf concepts (along with their relations) for validation and testing.[3] The remaining FoS taxonomy is then treated as the input existing taxonomy. We refer to this dataset as **MAG-Full**. Based on MAG-Full, we construct another dataset focusing on the computer science domain. Specifically, we first select a subgraph consisting of all descendants of "computer science" node and then mask 10% of leaf concepts in this subgraph for validation and another 10% of leaf nodes for testing. We name this dataset as **MAG-CS**.

Compared Methods. We compare the following methods:

- **Closest-Parent**: A rule-based method which first scores each candidate position in the existing taxonomy based on its cosine distance to the query concept between their initial embedding, and then ranks all positions using this score. The position with the smallest distance is chosen to be query concept's parent.
- **Closest-Neighbor**: Another rule-based method that scores each position based on its distance to the query concept plus the average distance between its children and the query.
- **dist-XGBoost**: A self-supervised boosting method that works directly on 39 manually-designed features generated using initial node embeddings without any embedding transformation. We input these features into XGBoost [5], a tree-based boosting model, to predict the matching score between a query concept and a candidate position.
- **ParentMLP**: A self-supervised method that first concatenates the query concept embedding with the candidate position embedding and then feeds them into a Multi-Layer Perceptron (MLP) for prediction.
- **DeepSetMLP**: Another self-supervised method that extends ParentMLP by adding information of candidate position's children nodes. Specifically, we first use DeepSet architecture [38] to generate the representation of the children node set and then concatenate it with query & candidate position representations before the final MLP module.
- **TaxoExpan**: Our proposed framework using position-enhanced GAT (PGAT) as graph propagation module and weighted mean readout (WMR) for graph readout. We learn this model using our proposed InfoNCE loss.

Evaluation Metrics. As our model returns a rank list of all candidate parents for each input query concept, we evaluate its performance using three ranking-based metrics.

- **Mean Rank (MR)** measures the average rank position of a query concept's true parent among all candidates. For queries with multiple parents, we first calculate the rank posi-

---

[3] Here we mask only leaves because if we remove intermediate nodes, we have to remove their descendants from the candidate parent pool, which causes different masked nodes (as testing query concepts) having different candidate pools.

tion of each individual parent and then take the average of all rank positions. Smaller MR value indicates better model performance.

- **Hit@k** is the number of query concepts whose parent is ranked in the top k positions, divided by the total number of queries.
- **Mean Reciprocal Rank (MRR)** calculates the reciprocal rank of a query concept's true parent. We follow [33] and use a scaled version of MRR in the below equation:

$$\text{MRR} = \frac{1}{|C|} \sum_{c \in C} \frac{1}{|parent(c)|} \sum_{i \in parent(c)} \frac{1}{\lceil R_{i,c}/10 \rceil}, \tag{4.18}$$

where $parent(c)$ represents the parent node set of the query concept c, and $R_{i,c}$ is the rank position of query concept c's true parent i. We scale the original MRR by a factor 10 in order to amplify the performance gap between different methods.

Implementation Details. For a fair comparison, we use the same 250-dimension embeddings across all compared methods. We use Google's original word2vec implementation[4] for learning embeddings and employ gensim[5] to load trained embeddings for calculating term distances in Closest-Parent, Closest-Neighbor, and dist-XGBoost methods. For the other three methods, we implement them using PyTorch and DGL framework.[6] We tune hyperparameters in all self-supervised methods on the masked validation set. For TaxoExpan, we use a two-layer position-enhanced GAT where the first layer has four attention heads (of size 250) and the second layer has one attention head (of size 500). For both layers, we use 50-dimension position embeddings and apply dropout with rate 0.1 on the input feature vectors. We use Adam optimizer with initial learning rate 0.001 and ReduceLROnPlateau scheduler[7] with three patience epochs. We discuss the influence of these hyper-parameters in the next subsection.

4.4.1.2 Experiment Results on MAG Datasets

We present the experiment results in the following aspects.

Overall Performance. Table 4.2 presents the results of all compared methods. First, we find that Closest-Neighbor method clearly outperforms Closest-Parent method. Also, the DeepSetMLP method is much better than ParentMLP. This demonstrates the effectiveness of modeling local structure information. Second, we compare dist-XGBoost method with Closest-Neighbor and show that self-supervision indeed helps us to learn an effective way to combine various neighbor distance information. All four self-supervised methods outper-

---

[4] https://github.com/tmikolov/word2vec.

[5] https://github.com/RaRe-Technologies/gensim.

[6] https://github.com/dmlc/dgl.

[7] https://pytorch.org/docs/stable/optim.html#torch.optim.lr_scheduler.ReduceLROnPlateau.

Table 4.2 Overall results on MAG-CS and MAG-Full datasets. We run all methods three times and report the averaged result with the the best two models highlighted under each metric

Method	MAG-CS				MAG-Full			
	MR	Hit@1	Hit@3	MRR	MR	Hit@1	Hit@3	MRR
Closest-Parent	1327.16	0.0531	0.0986	0.2691	14355.5	0.0360	0.0728	0.1897
Closest-Neighbor	382.07	0.1085	0.2000	0.3987	4160.8	0.0221	0.0419	0.1405
dist-XGBoost	136.86	0.1903	0.3483	0.6618	**426.70**	**0.1498**	**0.3046**	0.5621
ParentMLP	**114.79**	0.0729	0.2656	0.6454	457.14	0.098	0.1928	0.4950
DeepSetMLP	115.26	**0.1988**	**0.3581**	**0.6653**	444.83	0.1461	0.2971	**0.6392**
TaxoExpan	**80.33**	**0.2121**	**0.3823**	**0.6929**	**341.31**	**0.1523**	**0.3087**	**0.6453**

form rule-based methods. Finally, our proposed TaxoExpan has the overall best performance across all the metrics and defeats the second best method by a large margin.

Ablation Analysis of Model Architectures. TaxoExpan contains three key components: a graph propagation module, a graph readout module, and a matching model. Here, we study how different choices of these components affect the performance of TaxoExpan. Table 4.3 lists the results and the first column contains the index of each model invariant.

First, we analyze graph propagation module by using simple average scheme for graph readout and MLP for matching. By comparing model 1 to model 3 and model 2 to model 4, we can see that graph attention architecture (GAT) is better than graph convolution architecture (GCN). Furthermore, the position-enhanced variants clearly outperform their non-position counterparts (model 3 vs. model 1 and model 4 vs. model 2). This illustrates the efficacy of the position embeddings in the graph propagation module.

Second, we study graph readout module by fixing the graph propagation module to be the best two variants among models 1–4. We can see both model 5 & 6 outperform model 3 and model 7 & 8 outperform model 4. This signifies that the position information also helps in the graph readout module. However, the best strategy of incorporating position information depends on the graph propagation module. The concatenation readout scheme works better for PGCN while the weighted mean readout is better for PGAT. One possible explanation is that the concatenation readout leads to more parameters in matching model and as PGAT itself has more parameters than PGCN, further introducing more parameters in PGAT may cause the model to be overfitted.

Finally, we examine the effectiveness of different matching models. We replace the MLP in models 5–8 with LBM to create model variants 9–12. We can clearly see that LBM works better than MLP. It could be that LBM better captures the interaction between the query representation and the final anchor representation.

Table 4.3 Ablation analysis of model architectures on MAG-CS dataset. We assign an index to each model variant (shown in the first column). All models are run three times with their averaged scores reported

Ind	Graph propagate	Graph readout	Matching	MR	Hit@1	1Hit@3	1MRR
1	GCN	Mean	MLP	167.82	0.1581	0.2964	0.6002
2	GAT	Mean	MLP	131.46	0.1584	0.3192	0.6409
3	PGCN	Mean	MLP	148.54	0.1809	0.3015	0.6255
4	**PGAT**	**Mean**	**MLP**	**100.80**	**0.1896**	**0.3304**	**0.6525**
5	PGCN	WMR	MLP	144.81	0.1798	0.3014	0.6309
6	PGCN	CR	MLP	135.89	0.1902	0.3118	0.6348
7	**PGAT**	**WMR**	**MLP**	**92.62**	**0.1945**	**0.3584**	**0.6619**
8	PGAT	CR	MLP	95.84	0.1897	0.3512	0.6596
9	PGCN	WMR	LBM	139.41	0.1829	0.3370	0.6642
10	PGCN	CR	LBM	130.12	0.1934	0.3462	0.6776
11	**PGAT**	**WMR**	**LBM**	**80.33**	**0.2121**	**0.3823**	**0.6929**
12	PGAT	CR	LBM	84.40	0.2089	0.3813	0.6894

Ablation Analysis of Training Schemes. In this subsection, we evaluate the effectiveness of our proposed training scheme. In this study, we first group a set of positive and negative $\langle query, anchor \rangle$ pairs into *one single* training instance and learn the model using InfoNCE loss (c.f. Eq. (4.14)). An alternative is to treat these pairs as different instances and train the model using standard binary cross entropy (BCE) loss. Under this training scheme, we formulate our problem as a binary classification task. We compare these two training schemes for the top 4 best models in Table 4.3 (i.e., model 7, 8, 11, and 12). Results are shown in Fig. 4.5. Our proposed training scheme with InfoNCE loss is overall much better, it beats the BCE loss scheme on 14 out of total 16 cases. One reason is that BCE loss is very sensitive to the noises in the generated self-supervision data while InfoNCE loss is more robust to such label noise. Furthermore, we find that LBM matching can benefit more from our training scheme with InfoNCE loss—with larger margin on all 8 cases, compared with the simple MLP matching.

Hyper-Parameter Sensitivity Analysis. We analyze how some hyper-parameters in Taxo-Expan affect the performance in Fig. 4.6. First, we find that choosing an approximate position embedding dimension is important. The model performance increases as this dimensionality increases until it reaches about 50. When we further increase position embedding dimension, the model will overfit and the performance decreases. Second, we study the effect of negative sampling ratio N. As shown in Fig. 4.6, the model performance first increases as N increases until it reaches about 30 and then becomes stable. Finally, we examine two hyper-parameters controlling the model complexity: the number of heads in PGAT and the

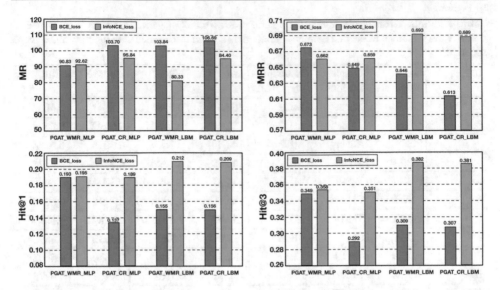

Fig. 4.5 Ablation analysis of training schemes on MAG-CS dataset. We compare models trained using Binary Cross Entropy (BCE) loss with those trained using InfoNCE loss

final graph embedding dimension. We observe that the best model performance is reached when the number of attention heads falls in range 3–5 and the graph embedding dimension is set to 500. Too many attention heads or too large graph embedding dimension will lead to overfit and performance degradation.

Efficiency and Scalability. We further analyze the scalability of TaxoExpan and its efficiency during model inference stage. Fig. 4.7 (left) tests the model scalability by running on MAG-CS dataset sampled using different ratios. The training time (of 20 epochs) are measured on one single K80 GPU. TaxoExpan demonstrates a linear runtime trend, which validates our complexity analysis in Sect. 4.3.4. Second, Fig. 4.7 (right) shows that TaxoExpan is very efficient during model inference stage. Using GPU, TaxoExpan takes <30 s to predict the anchor positions for all 2450 new query concepts.

Case Studies. Figures 4.8 and 4.9 shows some outputs of TaxoExpan on both MAG-CS and MAG-Full datasets. On MAG-CS dataset, we can see that over 20% of queries have their true parents correctly ranked at the first position and <1.5% queries have their "true" parents ranked outside of top 1000 positions. Among these 1.5% significantly wrong queries, we find some of them actually have incorrect existing parents. For example, the concept "*boils and carbuncles*", which is a disease entity, is mistakenly put under parent node "*dataset*". We also observe two common mistake patterns. The first type of mistakes is caused by term ambiguity. For instance, the term "*java*" in concept "*java apple*" refers to an island in Indonesia where fruit apple is produced, rather than a programming language used in Apple company. The second type of mistakes results from term granularity. For example, TaxoExpan outputs the two most likely parent nodes of concept "*captcha*" are "*artificial intelligence*" and "*computer security*".

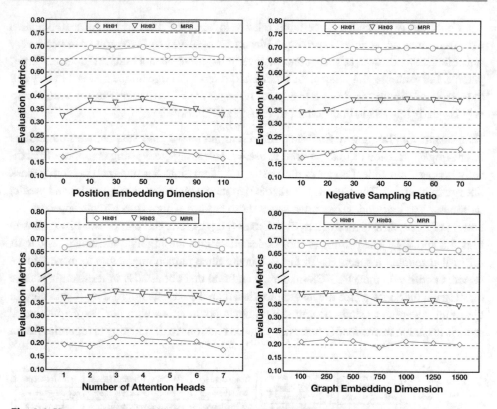

Fig. 4.6 Hyper-parameter sensitivity analysis on MAG-CS dataset. We use PGAT for graph propagation, WMR for graph readout, and LBM for query-graph matching. Model is trained using InfoNCE loss

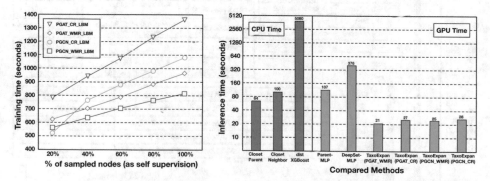

Fig. 4.7 (Left) Training time of 20 epochs on GPU with respect to % of sampled nodes in the existing taxonomy. (Right) Inference time of all 2450 queries in MAG-CS dataset. Note here y-axis is in logarithm scale

Although these two concepts are certainly relevant to "*captcha*", they are too general compared to its true parent node "*internet privacy*". Finally, we observe that TaxoExpan can return very sensible anchor positions of query concepts, even though they are not exactly the current "true" parents. For example, the concept "*medline plus*" refers to a large online medical library and thus is related to both "*world wide web*" and "*library science*". Also, the concept "*email hacking*" is clearly relevant to both "*internet privacy*" and "*hacker*".

TaxoExpan for Taxonomy Self-cleaning. From the above case studies, we find another interesting application of TaxoExpan is to use it for cleaning the existing taxonomy. Specifically, we partition all leaf nodes of the existing taxonomy into 5 groups and randomly mask one group of nodes. Then, we train a TaxoExpan model on the remaining nodes and predict on the masked leaf nodes. Next, we select those entities whose true parents appear at the bottom of the rank lists returned by TaxoExpan (i.e., the long-tail part of two histograms in Figs. 4.8 and 4.9). The parents of those selected entities are highly questionable and calls for further manual inspections. In fact, we invite three human annotators to inspect those selected entities on the MAG-CS taxonomy and find that about 30% of these entities have existing parent nodes which are less appropriate than the parents inferred by TaxoExpan. In the future, we plan to study how to equip human taxonomists with our TaxoExpan-based models to enable them identify potentially flawed relations in the existing taxonomies faster.

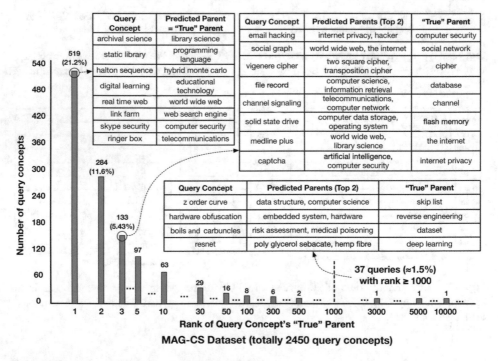

Fig. 4.8 Example output of TaxoExpan on MAG-CS dataset

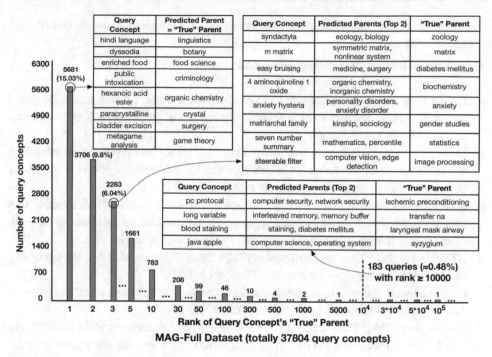

Fig. 4.9 Example output of TaxoExpan on MAG-Full dataset

4.4.2 Experiments on SemEval Dataset

4.4.2.1 Experiment Settings on SemEval Dataset

Datasets. Besides using the MAG datasets, we also evaluate TaxoExpan using SemEval Task 14 Benchmark dataset[8] [13] which includes WordNet 3.0 as the existing taxonomy and additional 1000 domain-specific concepts with manual labels, split into 400 training concepts and 600 testing concepts. Each concept is either a verb or a noun and has a textual definition of a few sentences. The original task goal is to enrich the taxonomy by performing two actions for each new concept: (1) *attach*, where a new concept is treated as a new synset and is attached as a hyponym of one existing synset in WordNet, and (2) *merge*, where a new concept is merged into an existing synset. However, previous state-of-the-art methods [13, 22, 28], including the winning solution, are only performing the *attach* operation. In this work, we also follow this convention and attach each new concept to the top-ranked synset in the WordNet. Finally, we obtain the initial feature vectors (for both new concepts and existing words in the WordNet) using pre-trained subword-aware fasttext embeddings.[9] For each concept, we generate its definition embedding and name embedding by averaging the embedding of each token in its textual definition and name string, correspondingly. Then, we

---

[8] http://alt.qcri.org/semeval2016/task14/.

[9] We use the wiki-news-300d-1M-subword.vec.zip version on fastText official website.

sum the definition and name embeddings of a concept and use them as the initial embeddings for the TaxoExpan model.

Compared Methods. We compare TaxoExpan with the following methods:

- **FWFS** [13]: The original baseline in Task 14. Given a concept c with its definition d_c, this method picks the first word w in d_c that has the same part of speech as c and treats this word as the parent node of c.
- **MSejrKU** [22]: The winning solution of Task 14. This method leverages distributional and syntactic features to train a SVM classifier which is then used to predict the goodness of fit for a new concept with an existing synset in WordNet.
- **ETF** [28]: The current state-of-the-art method that learns a LambdaMART model with 15 manually designed features, including topological features from the taxonomy's graph structure and semantic features from corpus and Bing search results.
- **ETF-FWFS** [28]: The ensemble model of FWFS and ETF, which adds the FWFS property as a binary feature into the LambdaMART model in ETF.
- **dist-XGBoost**: The same tree boosting model described in the previous subsection.
- **TaxoExpan**: Our proposed taxonomy expansion framework.
- **TaxoExpan-FWFS**: Similar to ETF-FWFS, this is the ensemble model of FWFS and TaxoExpan. We treat the FWFS heuristic as a binary feature and add it into the final matching module.

For all previous methods, we directly report their best performance in the literature. For the remaining methods, we tune them following the same procedure as described before.

Evaluation Metrics. We use the three official metrics defined in original SemEval Task 14 for evaluation:

- **Accuracy (Wu&P)** is the semantic similarity between a predicted parent node x_p and the true parent x_t, calculated as $Wu\&P(x_p, x_t) = \frac{2 \cdot depth_{LCA(x_p, x_t)}}{depth_{x_p} + depth_{x_t}}$, where $depth_x$ is the depth of node x is the WordNet taxonomy and $LCA(x_p, x_t)$ represents the Least Common Ancestor of x_p and x_t.
- **Recall** is the percentage of concepts for which an attached parent is predicted.[10]
- **F1** is the harmonic mean of Wu&P accuracy and recall.

4.4.2.2 Experiment Results on SemEval Datasets

Table 4.4 shows the experiment results on SemEval dataset. We can see that both dist-XGBoost and TaxoExpan methods can outperform the previous winning system of this task

---

[10] This metric is used because the original task allows a model to decline to place new concepts in order to avoid making placements with low confidence.

Table 4.4 Model performance on SemEval dataset. TaxoExpan versus all previous state-of-the-art methods. We report the best performance of all existing methods in the literature

Method	Wu&P	Recall	F1
MSejrKU [22]	0.523	0.973	0.680
FWFS [13]	0.514	1.000	0.679
ETF [28]	0.473	1.000	0.642
ETF-FWFS [28]	0.562	1.000	0.720
dist-XGBoost	0.528	1.000	0.691
TaxoExpan	0.543	1.000	0.704
TaxoExpan-FWFS	0.566	1.000	0.723

(i.e., MSejrKU) and the baseline ETF. In addition, we can see the FWFS heuristic is indeed very powerful for this dataset and incorporating it as a strong feature can significantly boost the performance. However, this feature requires human-labeled definition sentences and thus can not be easily generalized to taxonomies other than WordNet. Finally, we show that TaxoExpan-FWFS can achieve the new state-of-the-art performance on this dataset.

4.5   Extensions of TaxoExpan

TaxoExpan demonstrates an effective self-supervised framework for taxonomy expansion. However, it focuses only on the first-order parent-children relation in the self-supervision generation step and identifies only the parent concepts of new concepts. Together with our collaborators, we further extend TaxoExpan by (1) incorporating more fine-grained self-supervision tasks [37], (2) modeling relations among news concepts [26], and (3) identifying potential children concepts [42].

4.5.1   Incorporating More Fine-Grained Self-supervision Tasks

The essence of the self supervision generation step in TaxoExpan is to identify if a given ⟨query concept, anchor concept⟩ pair is positive or not. Although TaxoExpan incorporates a graph neural network to capture the structure information surrounding anchor concept, the self supervision task is still focusing on the first-order parent-children relation. To address this limitation, we propose STEAM which stands for "Self-supervised Taxonomy ExpAnsion with Mini-Paths".

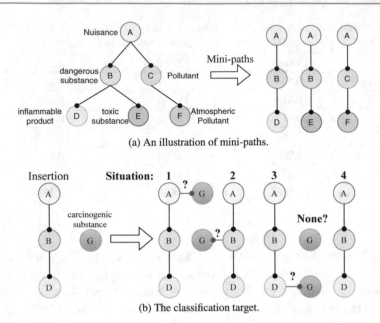

(a) An illustration of mini-paths.

(b) The classification target.

Fig. 4.10 An illustration of the proposed mini-paths and the mini-path-based node attachment task

These "mini-paths" are fixed-length paths sampled from the existing taxonomy, as shown in Fig. 4.10a. Each node in the mini-path serves as a potential parent node for a new query concept and the self-supervised learning task is to pinpoint which node in this mini-path is the correct parent. As shown in Figure 4.10b, given a length-3 mini-path "A–B–D" and a query concept "G", STEAM predicts the probabilities of the query concept "G" being attached to the three terms, or none of them. Compared with the simple task of binary hypernymy classification, matching query concepts with mini-paths has two major advantages. First, when attaching a query concept, considering the terms in mini-path provides richer information for query attachment than considering each term separately. Second, compared with the binary classification, this task is more challenging and thus the matching module needs to judge not only whether query concept should be matched to the whole mini-path but also which specific position to attach. Learning from this more challenging task allows STEAM to better leverage the structural information in the existing taxonomy.

We evaluate the performance on STEAM using three datasets in SemEval 2016 [4] which correspond to three human-curated concept taxonomies from different domains: environment (EN), science (SCI), and food (Food). For each taxonomy, we start from the root term and randomly grow in a top-down manner until 80% terms are covered. We use the randomly grown taxonomies as seed taxonomies for self-supervised learning, and the rest 20% terms as our test data.

We compare STEAM with TaxoExpan as well as three more baseline methods: (1) **TAXI** [20] is a taxonomy induction method that reached the first place in the SemEval 2016

Table 4.5 Comparison of STEAM against the baseline methods on the three datasets (in %). To reduce randomness, we ran all methods for three times and report the average performance. TAXI outputs an entire taxonomy instead of ranking lists, so we are unable to obtain its MRRs

Dataset	Environment			Science			Food		
Metric	Acc	MRR	Wu&P	Acc	MRR	Wu&P	Acc	MRR	Wu&P
BERT+MLP	11.1	21.5	47.9	11.5	15.7	43.6	10.5	14.9	47.0
TAXI	16.7	–	44.7	13.0	–	32.9	18.2	–	39.2
HypeNet	16.7	23.7	55.8	15.4	22.6	50.7	20.5	27.3	63.2
TaxoExpan	11.1	32.3	54.8	27.8	44.8	57.6	27.6	40.5	54.2
STEAM	**36.1**	**46.9**	**69.6**	**36.5**	**48.3**	**68.2**	**34.2**	**43.4**	**67.0**

task. It first extracts hypernym pairs based on substrings and lexico-syntactic patterns with domain-specific corpora and then organizes these terms into a taxonomy, (2) **HypeNet** [24] is a strong hypernym extraction method, which uses an LSTM model to jointly model the distributional and relational information between term pairs, and (3) **BERT+MLP** is a distributional method for hypernym detection based on pre-trained BERT embeddings. For each term pair, it first obtains term embeddings from a pre-trained BERT model and then feeds them into a Multi-layer Perceptron to predict whether they have the hypernymy relationship.[11]

As shown in Table 4.5, STEAM consistently outperforms all the baselines by large margins on the three datasets. In particular, STEAM improves the performance of the state-of-the-art TaxoExpan model by 11.6, 7.0 and 9.4% for Acc, MRR and Wu&P on average. Such improvements are mainly due to the mini-path-based prediction and the multi-view co-training designs in TaxoExpan.

4.5.2 Identifying Potential Children Concepts

One limitation of TaxoExpan and STEAM is that they assume new concepts can only be added into the existing taxonomy as hyponyms (i.e., leaf nodes[12]) However, such an assumption is inappropriate in some real applications. For example, in Fig. 4.11, the term "*Smart Phone*" is invented much later than term "*CPU*", which means that when "*Smart Phone*" emerges, "*CPU*" already exists in taxonomy In this case, it is inappropriate to add "*Smart Phone*" into taxonomy as leaf node because "*CPU*" is a hyponym of "*Smart Phone*".

To address this limitation, we define and investigate a new *taxonomy completion* task *without* the strong "hyponym-only" assumption. Formally, given an existing taxonomy and

---

[11] For combining term embeddings, we experiment with CONCAT, DIFFERENCE, and SUM as different fusing functions and report the best performance.

[12] Nodes with zero out-degree in a directed acyclic graph.

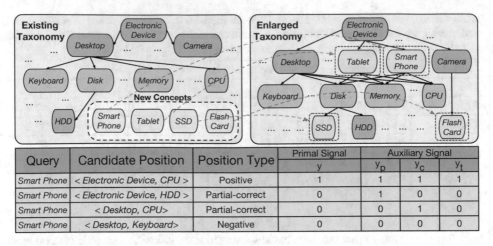

Query	Candidate Position	Position Type	Primal Signal	Auxiliary Signal		
			y	y_p	y_c	y_t
Smart Phone	*< Electronic Device, CPU >*	Positive	1	1	1	1
Smart Phone	*< Electronic Device, HDD >*	Partial-correct	0	1	0	0
Smart Phone	*< Desktop, CPU>*	Partial-correct	0	0	1	0
Smart Phone	*< Desktop, Keyboard>*	Negative	0	0	0	0

Fig. 4.11 An example of completing one "*Electronic Device*" taxonomy. The table illustrates different types of candidate positions for a given query "*Smart Phone*"

a set of new concepts, we aim to automatically complete the taxonomy to incorporate these new concepts by discovering the most likely ⟨hypernym, hyponym⟩ pairs of each new concept. For instance, in Fig. 4.11, one of the most likely candidate pairs for "*Smart Phone*" is ⟨"*Electronic Device*", "*CPU*"⟩. This formulation leads to a novel *one-to-pair* matching problem different from the previous *one-to-one* setting in TaxoExpan and STEAM. Solving this taxonomy completion task enables us to identify both the parent (hypernym) of the query concept and its the children (hyponyms). Note that the hypernym/hyponym concept within the candidate ⟨hypernym, hyponym⟩ pair could be a "pseudo concept" in case there is no appropriate one for a given query concept. We can easily see that the taxonomy expansion task is a special case of taxonomy completion task when the hyponym concepts are always "pseudo concept".

We propose a novel **Triplet Matching Network** (TMN) [42] to address the taxonomy completion task. One challenge of this task is that the induced one-to-pair matching problem results in a special type of negative candidate ⟨hypernym, hyponym⟩ pairs which we called *partially-correct negative candidates*. For a given query concept n_q, a candidate pair of existing concepts ⟨n_p, n_c⟩ is *positive* if $n_p(n_c)$ is the true hypernym (hyponym) of n_q and *negative* otherwise. Then, a candidate pair ⟨n_p, n_c⟩ is partially-correct negative if either n_p is true hypernym but n_c is not true hyponym or vice versa (as shown in Fig. 4.11). These partially-correct negative candidate typically have high correlations with positive pairs which makes model struggle to distinguish one from another To solve this challenge, TMN jointly learns a scoring function to output the matching score of a ⟨query, hypernym, hyponym⟩ triplet and leverages *auxiliary signals* to help distinguish positive pairs from partially-correct negative ones These auxiliary signals are binary signals indicating whether one component within the pair is positive or not, in contrast to binary *primal signals* that

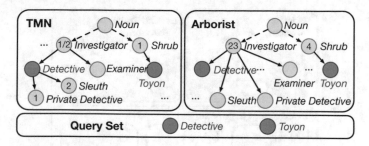

Fig. 4.12 Case study. The grey concepts are existing concepts while the red ones are queries needed to be inserted. The dash lines mean an omitting of some internal nodes and the solid lines indicate real edges in taxonomy. The numbers inside nodes are the ranking position output by the model. We can see TMN recovers the true positions for both internal and leaf concepts

reveal holistically whether a triplet is positive or not. TMN consists of multiple *auxiliary scorers* that learn different auxiliary signals via corresponding auxiliary loss and one *primal scorer* that aggregates internal feature representations of auxiliary scorers to output the final matching score. The auxiliary and primal scorers are jointly trained in an auxiliary learning framework which encourages the model to learn meaningful feature representations for the primal scorer to discriminate between positive, negative, and partially-correct negative candidates.

We illustrate the power of TMN via two real query concepts *"Detective"* and *"Toyon"* of **WordNet-Noun** in Fig. 4.12. For internal concept *"Detective"*, TMN ranks the true positions ⟨*"Investigator"*, *"Private Detective"*⟩ at top 1 and ⟨*"Investigator"*, *"Sleuth"*⟩ at top 2, while Arborist [18], a popular taxonomy expansion method, can only rank the true parent *"Investigator"* at top 23. For leaf concept *"Toyon"*, TMN recovers its true parent *"Shrub"* but Arborist ranks *"Shrub"* at top 5. We can see that TMN works better than baseline in terms of recovering true positions.

4.5.3 Modeling Relations Among News Concepts

With rapid growth of human knowledge, new concepts are emerging in batches instead of one at a time. All the studies we have discussed in previous sections assume these concepts are independent, which implies the insertion order of them into the existing taxonomy is not important. However, these new concepts typically have inter-relations and thus the order of inserting them into the existing taxonomy is critical. Suppose that a hyponym concept is inserted before its hypernyms, existing approaches can hardly recover the ground truth hypernym-hyponym structure. For example, considering a hypernym-hyponym pair of concepts ("geometry", "rectangle"), if "rectangle" is first inserted into the existing taxonomy, then when processing "geometry", we can only insert it as a hyponym of "rectangle", which is incorrect. Figure 4.13 (Bottom) illustrates how inserting order determines the quality of

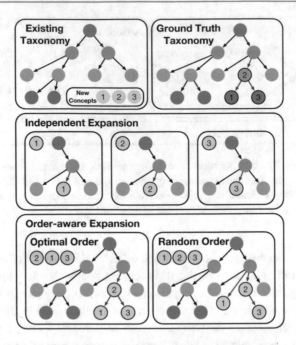

Fig. 4.13 An illustrative example of the taxonomy expansion task. (Top-left): an existing taxonomy with three new concepts. (Top-right) the ground truth taxonomy we want to discover. (Middle): independent expansion, where the taxonomy expansion task is divided into 3 independent tasks. Each new concepts can only be inserted as children of existing taxonomy concepts. No hypernym-hyponym relation among new concepts can be discovered. (Bottom): order-aware expansion, we illustrate the expanded taxonomy with optimal and random order of concepts to be inserted in order-aware taxonomy expansion

expanded taxonomy. For an optimal taxonomy expansion model which always outputs the "correct" hypernym concept, if the inserting order is sub-optimal (1-3-2 or 3-1-2), it can never recover ground truth taxonomy. To bypass such a problem, TaxoExpan and STEAM divide the task of inserting k new concepts into k *independent* tasks whose input new concept set contains only one element. As illustrated in the "Independent Insertion" case in Fig. 4.13 (Middle), this strategy ignores the potential dependencies of new concepts and restricts one new concept to be inserted underneath another new concept.

To address this issue, we instead study the *order-aware* taxonomy expansion task as shown in Fig. 4.13 (Bottom). We aim to discover the appropriate order of insertion and iteratively insert each individual new concept into the existing taxonomy. We propose a novel order-aware taxonomy expansion model TaxoOrder [26]. The key idea of TaxoOrder is to determine an optimal insertion order of new concepts. Within this order, hypernyms are ranked in front of hyponyms and thus when hyponyms are inserted, their potential parent nodes are already in the partially expanded taxonomy. TaxoOrder first learns the hypernym-hyponym relations from the existing taxonomy. Then, it utilizes the learned model together with heuristic patterns to generate pseudo-edges of new concepts, which carries the relative hypernym-hyponym information between each concept pair. Finally, we decide the insertion order by topological sorting on the DAG constructed from pseudo-edges. TaxoOrder can be combined any existing taxonomy expansion system (e.g., TaxoExpan, STEAM, Arborist) for improving the performance of taxonomy expansion.

We evaluate TaxoOrder on the modified MAG-CS dataset. Specifically, we only mask leaf nodes for validation. For testing, if some node concept c is sampled, we mask the whole DAG rooted at c to construct the testing set. Besides the metrics discussed in Sect. 4.4.1.1, we also use the following metrics: (1) **Error Node Count (ENC)** which is the number of query concepts whose parent is not present in the existing taxonomy when it is inserted and (2) **Edge F1** which is calculated based on the expanded taxonomy ($\mathcal{T}_{pred}$) edges and ground truth taxonomy ($\mathcal{T}_{gt}$) edges.

$$\mathbf{P} = \frac{|\mathcal{E}(\mathcal{T}_{pred}) \cap \mathcal{E}(\mathcal{T}_{gt})|}{|\mathcal{E}(\mathcal{T}_{pred})|}, \quad \mathbf{R} = \frac{|\mathcal{E}(\mathcal{T}_{pred}) \cap \mathcal{E}(\mathcal{T}_{gt})|}{|\mathcal{E}(\mathcal{T}_{gt})|}, \quad \mathbf{Edge\ F1} = \frac{2 * \mathbf{P} * \mathbf{R}}{\mathbf{P} + \mathbf{R}}.$$

We compare our TaxoOrder with several baselines. All these method are used to determine the inserting order for new concepts. We examine the effectiveness of these methods with two taxonomy expansion modules: TaxoExpan and Arborist. Note that taxonomy expansion module is independent to our concept sorting algorithm, which makes our TaxoOrder compatible with other taxonomy expansion work.

1. **Random**: This method simply inserts new concepts in random order which applies the taxonomy expansion module directly to the new task.
2. **Affinity**: The taxonomy expansion module typically assigns an affinity score for each candidate ⟨query concept, anchor concept⟩ pair. Such affinity scores can be interpreted as the level of prediction confidence. Thus, this **Affinity** method inserts new concepts based on the affinity scores from the taxonomy expansion module. In other words, the new concept node with the highest affinity score with the existing taxonomy node gets inserted first.
3. **TaxoOrder**: Our proposed TaxoOrder method.

Table 4.6 shows the overall result of all compared methods. First, by comparing the correlation between ENC and other metrics, it can be concluded that the taxonomy expansion performance is highly sensitive to the inserting order: the lower the ENC, the higher the expansion performance can achieve. And the ground truth inserting order gives the best expansion performance since its ENC is zero. Among all the methods compared in both expansion modules in experiments, the TaxoOrder outperforms the baseline methods and is distinguished out by a large margin. Even compared with the ground truth order, the TaxoOrder performs relatively well in terms of most of the evaluation metrics. Comparing Affinity with Random, the ENC has dropped a lot, but in the expansion quality evaluation, there is only a small improvement from Random to Affinity. That is because the Affinity

Table 4.6 Overall results on MAG-CS datasets with TaxoExpan and Arborist. Note that smaller ENC indicates better model performance. For all other metrics, larger values indicate better performance. We highlight the best two models in terms of the performance under each metric

Methods	ENC	MRR	Hit@1	Hit@3	Edge F1
TaxoExpan + Ground Truth	**0**	**0.2702**	**0.1934**	**0.2576**	**0.9026**
TaxoExpan + Random	1208	0.2113	0.1450	0.2109	0.8985
TaxoExpan + Affinity	966	0.2169	0.1498	0.2157	0.8989
TaxoExpan + TaxoOrder	**437**	**0.2595**	**0.1782**	**0.2534**	**0.9013**
Arborist + Ground Truth	**0**	**0.2272**	**0.2024**	**0.2619**	**0.9034**
Arborist + Random	1208	0.1877	0.1450	0.2135	0.8994
Arborist + Affinity	966	0.1919	0.1482	0.2183	0.8995
Arborist + TaxoOrder	**437**	**0.2258**	**0.1780**	**0.2584**	**0.9013**

score is not tailored to this sorting task. Although it represents the hypernym-hyponym relation to some extent, the lack of learning process on this sorting task limits its capability to get better expansion results.

4.6 Summary

This chapter studies the problem of concept taxonomy enrichment when no human labeled supervision data are given. We propose a novel TaxoExpan framework which generates self-supervision data from the existing taxonomy and learns a position-enhanced GNN model for expansion. To make the best use of self-supervision data, we design a noise-robust objective for effective model training. Extensive experiments demonstrate the effectiveness and efficiency of TaxoExpan on three taxonomies from different domains. Moreover, we introduce a few ways to extend TaxoExpan by incorporating more fine-grained self-supervision tasks, identifying new concepts' children concepts, and modeling relations among news concepts.

Interesting future work includes modeling inter-dependency among new concepts, leveraging current method to cleaning the input existing taxonomy, and incorporating feedbacks from downstream applications (e.g., search & recommendation) to generate more diverse supervision signals for expanding the taxonomy.

References

1. Aly, R., Acharya, S., Ossa, A., Köhn, A., Biemann, C., Panchenko, A.: Every child should have parents: a taxonomy refinement algorithm based on hyperbolic term embeddings. In: Proceedings of the 57th Annual Meeting of the Association for Computational Linguistics (2019)
2. Anke, L.E., Camacho-Collados, J., Rodríguez-Fernández, S., Saggion, H., Wanner, L.: Extending WordNet with fine-grained collocational information via supervised distributional learning. In: Proceedings of the 26th International Conference on Computational Linguistics (2016)
3. Bentivogli, L., Bocco, A.A., Pianta, E.: ArchiWordNet: Integrating WordNet with domain-specific knowledge. In: Proceedings of the 2nd International Global Wordnet Conference (2003)
4. Bordea, G., Lefever, E., Buitelaar, P.: Semeval-2016 Task 13: Taxonomy extraction evaluation (TExEval-2). In: Proceedings of the 10th International Workshop on Semantic Evaluation (2016)
5. Chen, T., Guestrin, C.: XGBoost: A scalable tree boosting system. In: Proceedings of the 22nd ACM SIGKDD International Conference on Knowledge Discovery and Data Mining (2016)
6. Chen, J.J., Ma, T., Xiao, C.: FastGCN: Fast learning with graph convolutional networks via importance sampling. In: Proceedings of the 6th International Conference on Learning Representations (2018)
7. Fellbaum, C., Hahn, U., Smith, B.D.: Towards new information resources for public health—from WordNet to MedicalWordNet. J. Biomed. Inform. (2006)
8. Hamilton, W.L., Ying, Z., Leskovec, J.: Inductive representation learning on large graphs. In: Proceedings of the 31st Conference on Neural Information Processing Systems (2017)
9. Hua, W., Wang, Z., Wang, H., Zheng, K., Zhou, X.: Understand short texts by harvesting and analyzing semantic knowledge. In: IEEE Transactions on Knowledge and Data Engineering (2017)
10. Huang, J., Ren, Z., Zhao, W.X., He, G., Wen, J.R., Dong, D.: Taxonomy-aware multi-hop reasoning networks for sequential recommendation. In: Proceedings of the 12ND ACM International Conference on Web Search and Data Mining (2019)
11. Jin, W., Barzilay, R., Jaakkola, T.S.: Junction tree variational autoencoder for molecular graph generation. In: Proceedings of the 35th International Conference on Machine Learning (2018)
12. Jurgens, D., Pilehvar, M.T.: Reserating the awesometastic: an automatic extension of the WordNet taxonomy for novel terms. In: Proceedings of the 2015 Conference of the North American Chapter of the Association for Computational Linguistics: Human Language Technologies (2015)
13. Jurgens, D., Pilehvar, M.T.: SemEval-2016 task 14: semantic taxonomy enrichment. In: Proceedings of the 10th International Workshop on Semantic Evaluation (2016)
14. Kipf, T.N., Welling, M.: Semi-supervised classification with graph convolutional networks. In: Procidings of the 5th International Conference on Learning Representations (2017)
15. Lee, J.B., Rossi, R.A., Kong, X.: Graph classification using structural attention. In: Proceedings of the 24th ACM SIGKDD International Conference on Knowledge Discovery and Data Mining (2018)
16. Li, Y., Vinyals, O., Dyer, C., Pascanu, R., Battaglia, P.W.: Learning deep generative models of graphs. In: Proceedings of the 6th International Conference on Learning Representations (2018)

17. Liu, B.W., Guo, W., Niu, D., Wang, C., Xu, S.Z., Lin, J., Lai, K., Xu, Y.W.: A user-centered concept mining system for query and document understanding at tencent. In: Proceedings of the 25th ACM SIGKDD International Conference on Knowledge Discovery and Data Mining (2019)
18. Manzoor, E., Li, R., Shrouty, D., Leskovec, J.: Expanding taxonomies with implicit edge semantics. In: Proceedings of The Web Conference 2020, pp. 2044–2054 (2020)
19. Oord, A.V.D., Li, Y., Vinyals, O.: Representation learning with contrastive predictive coding (2018). arXiv preprint arXiv:1807.03748
20. Panchenko, A., Faralli, S., Ruppert, E., Remus, S., Naets, H., Fairon, C., Ponzetto, S.P., Biemann, C.: TAXI at SemEval-2016 task 13: a taxonomy induction method based on lexico-syntactic patterns, substrings and focused crawling. In: Proceedings of the 10th International Workshop on Semantic Evaluation (SemEval-2016), pp. 1320–1327 (2016)
21. Plachouras, V., Petroni, F., Nugent, T., Leidner, J.L.: A comparison of two paraphrase models for taxonomy augmentation. In: Proceedings of the 2018 Conference of the North American Chapter of the Association for Computational Linguistics: Human Language Technologies (2018)
22. Schlichtkrull, M.S., Alonso, H.M.: MSejrKu at SemEval-2016 task 14: taxonomy enrichment by evidence ranking. In: Proceedings of the 10th International Workshop on Semantic Evaluation (2016)
23. Shen, Z., Ma, H., Wang, K.: A web-scale system for scientific knowledge exploration. In: Proceedings of the 56th Annual Meeting of the Association for Computational Linguistics (2018)
24. Shwartz, V., Goldberg, Y., Dagan, I.: Improving hypernymy detection with an integrated path-based and distributional method. In: Proceedings of the 54th Annual Meeting of the Association for Computational Linguistics, Volume 1: Long Papers, pp. 2389–2398 (2016)
25. Sinha, A., Shen, Z., Song, Y., Ma, H., Eide, D., Hsu, B.J.P., Wang, K.: An overview of Microsoft Academic Service (MAS) and applications. In: Proceedings of the 2015 International Conference on World Wide Web (2015)
26. Song, X., Shen, J., Zhang, J., Han, J.: Who should go first? A self-supervised concept sorting model for improving taxonomy expansion. In: Proceedings of the International Workshop on Self-supervised Learning for the Web (2021)
27. Toral, A., Muñoz, R., Monachini, M.: Named entity WordNet. In: Proceedings of the 6th International Conference on Language Resources and Evaluation (2008)
28. Vedula, N., Nicholson, P.K., Ajwani, D., Dutta, S., Sala, A., Parthasarathy, S.: Enriching taxonomies with functional domain knowledge. In: Proceedings of the 41st International ACM SIGIR Conference on Research and Development in Information Retrieval (2018)
29. Velickovic, P., Cucurull, G., Casanova, A., Romero, A., Liò, P., Bengio, Y.: Graph attention networks. In: Proceedings of the 6th International Conference on Learning Representations (2018)
30. Wang, J., Kang, C., Chang, Y., Han, J.: A hierarchical Dirichlet model for taxonomy expansion for search engines. In: Proceedings of the 23rd International Conference on World Wide Web (2014)
31. Wu, W., Li, H., Wang, H., Zhu, K.Q.: Probase: a probabilistic taxonomy for text understanding. In: Proceedings of the 2012 ACM SIGMOD International Conference on Management of Data (2012)
32. Yang, G.H.: Constructing task-specific taxonomies for document collection browsing. In: Proceedings of the 2012 Conference on Empirical Methods in Natural Language Processing (2012)
33. Ying, R., He, R., Chen, K., Eksombatchai, P., Hamilton, W.L., Leskovec, J.: Graph convolutional neural networks for web-scale recommender systems. In: Proceedings of the 24th ACM SIGKDD International Conference on Knowledge Discovery and Data Mining (2018)

34. Ying, Z., You, J., Morris, C., Ren, X., Hamilton, W.L., Leskovec, J.: Hierarchical graph representation learning with differentiable pooling. In: Proceedings of the 32nd Conference on Neural Information Processing Systems (2018)
35. You, J., Liu, B., Ying, Z., Pande, V.S., Leskovec, J.: Graph convolutional policy network for goal-directed molecular graph generation. In: Proceedings of the 32nd Conference on Neural Information Processing Systems (2018)
36. You, J., Ying, R., Leskovec, J.: Position-aware graph neural networks. In: Proceedings of the 36th International Conference on Machine Learning (2019)
37. Yu, Y., Li, Y., Shen, J., Feng, H., Sun, J., Zhang, C.: STEAM: Self-supervised taxonomy expansion with mini-paths. In: Proceedings of the 26th ACM SIGKDD International Conference on Knowledge Discovery and Data Mining (2020)
38. Zaheer, M., Kottur, S., Ravanbakhsh, S., Póczos, B., Salakhutdinov, R., Smola, A.J.: Deep sets. In: Proceedings of 31st Conference on Neural Information Processing Systems (2017)
39. Zeng, Q., Lin, J., Yu, W., Cleland-Huang, J., Jiang, M.: Enhancing taxonomy completion with concept generation via fusing relational representations. In: Proceedings of the 27th ACM SIGKDD International Conference on Knowledge Discovery and Data Mining (2021)
40. Zhang, Y., Ahmed, A., Josifovski, V., Smola, A.J.: Taxonomy discovery for personalized recommendation. In: Proceedings of the 7th ACM International Conference on Web Search and Data Mining (2014)
41. Zhang, M., Cui, Z., Neumann, M., Chen, Y.: An end-to-end deep learning architecture for graph classification. In: Proceedings of the 2018 AAAI Conference on Artificial Intelligence (2018)
42. Zhang, J., Song, X., Zeng, Y., Chen, J., Shen, J., Mao, Y., Li, L.: Taxonomy completion via triplet matching network. In: Proceedings of the 2021 AAAI Conference on Artificial Intelligence (2021)

Taxonomy-Guided Classification

5

5.1 Overview and Motivations

With taxonomy constructed and enriched on a domain-specific document collection, we can leverage it to enhance lots of downstream knowledge-centric applications. For example, we can use taxonomies to organize scientific literatures and support semantic literature retrieval. Similarly, we can categorize products based on a product category taxonomy and build a better search and recommender system. Within all those applications, we assume the text unit (either an entire document or an in-context text span) has already been associated with a set of classes in the corresponding taxonomy. Such assumption, however, may not hold in many real-world applications for which the taxonomy powered methods can not be applied. Therefore, to fully exploit the power of taxonomy, we need to study the hierarchical multi-label text classification (HMTC) problem which aims to assign each text document to a set of relevant classes from a class taxonomy.

Most existing methods address HMTC in a supervised fashion—they first ask humans to provide many labeled documents and then train a text classifier for prediction. Many classifiers have been developed with different deep learning architectures such as CNN [11], RNN [34], Attention Network [8], and achieved decent performance when trained on massive human-labeled documents. Despite such a success, people find that applying these methods to many real-world scenarios remains challenging as the human labeling process is often too time-consuming and expensive.

Recently, more studies have been developed to address text classification using smaller amount of labeled data. First, several semi-supervised methods [2, 7] propose to use abundant unlabeled documents to assist model training on labeled dataset. Although mitigating the human annotation burden, these methods still require a labeled dataset that covers all classes, which could be too expensive to obtain when we have a large number of classes in HMTC. Second, some weakly-supervised models exploit class indicative keywords [19, 20, 35] or

© The Author(s), under exclusive license to Springer Nature Switzerland AG 2022
J. Shen and J. Han, *Automated Taxonomy Discovery and Exploration*,
Synthesis Lectures on Data Mining and Knowledge Discovery,
https://doi.org/10.1007/978-3-031-11405-2_5

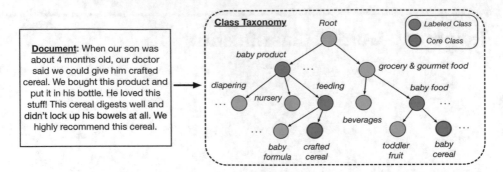

Fig. 5.1 An exemplar document tagged with five classes. Here, if we are able to pinpoint this document's most essential classes, *crafted cereal* and *baby cereal*, as core classes, we can check their ancestor classes in the taxonomy and recover all the true classes

class surface names [22, 29] to derive pseudo-labeled data for model training. Nevertheless, these models all assume each document has only one class and all class surface names (or class indicative keywords) must appear in the corpus, which are too restrictive for HMTC.

In this chapter, we study the problem of *weakly-supervised* hierarchical multi-label text classification where only class surface names, a class taxonomy, and an unlabeled corpus are available for model training. This setting is closer to how humans resolve the HMTC problem—we perform classification by understanding each class from its surface name rather than learning from labeled documents. We observe that when asked to assign multiple classes to a document, humans will first pinpoint most essential "core classes" and then check whether their ancestor classes in the taxonomy should also be tagged. Taking the document in Fig. 5.1 as an example, humans can quickly identify this review text is clearly about "*baby cereal*" and "*crafted cereal*", which are the core classes. After assigning these two most essential classes to the document, people continue to check the core classes' ancestor classes and find "*feeding*" as well as "*baby food*" should be tagged.

Motivated by the above human labeling process, we propose TaxoClass, a *weakly-supervised* HMTC framework including four major steps. First, we calculate the document-class similarity using a pre-trained textual entailment model [33]. Second, we identify each document's core classes by (1) selecting candidate core classes that are most similar to the document at each level in a top-down fashion, and (2) choosing ⟨document, candidate core class⟩ pairs that are salient across the whole unlabeled corpus. Third, we derive training data from document core classes and use them to train a text classifier. This classifier includes a document encoder based on pre-trained BERT [4], a class encoder capturing class taxonomy structure, and a text matching network computing the probability of a document being tagged with each class. Finally, we generalize this text classifier using multi-label self-training on all unlabeled documents.

To summarize, our major contributions are as follows:

- We propose a weakly-supervised framework TaxoClass that only requires class surface names to perform hierarchical multi-label text classification. To the best of our knowledge, TaxoClass is the first weakly-supervised HMTC method.
- We develop an unsupervised method to identify document core classes based on which a text classifier can be learned.
- We conduct extensive experiments on two real-world datasets to verify the effectiveness of TaxoClass.

We organize the rest of this chapter as follows. Section 5.2 discusses the related work. Section 5.3 presents our TaxoClass framework and we conduct experiments in Sect. 5.4. Finally, we conclude this chapter in Sect. 5.5.

5.2 Related Work

There are three major lines of work related to our study.

Weakly-supervised Text Classification. There exist some previous studies that leverage a few labeled documents or class-indicative keywords as weak supervision signals for text classification. A pioneering method is dataless classification [3, 27] which embeds documents and classes into the same semantic space of Wikipedia concepts and performs classification using the embedding similarity. After that, researchers extend this idea by mining concepts directly from the corpus rather than using the external Wikipedia [15, 16]. Along another line, Chen et al. and Li et al. propose to apply a seed-guided topic model to infer class-specific topics from class-indicative keywords and to predict document classes from posterior class-topic assignments. Compared with these methods, our TaxoClass framework neither restricts document and class embeddings to live in the same semantic space nor imposes strong statistical assumptions.

Recently, neural models are applied to weakly-supervised text classification. Researchers propose a pretrain-and-refine paradigm which first generates pseudo documents to pretrain a neural classifier and then refine this classifier via self-training [20, 21]. More studies [19, 22, 29] improve the above methods by introducing contextualized weak supervision and using a pre-trained language model to obtain better text representations. While achieving inspiring performance, these methods all assume each document has only one class and all class names (or class-indicative keywords) must appear in the corpus for pseudo training data generation. In this work, we relax these assumptions and develop a new method for weakly-supervised hierarchical multi-label text classification task.

Zero-shot Text Classification. Zero-shot text classification learns a text classifier based on training documents belonging to *seen* classes and applies the learned classifier to predict testing documents belonging to *unseen* classes. Nam et al. [23] jointly embed documents and classes into a shared semantic space where knowledge from seen classes can be transferred to unseen classes. Such an idea is further developed in [25, 28, 33] where external resources (e.g., knowledge graphs, natural language explanations of unseen classes, and open domain data) are introduced to help learn a better shared semantic space. Comparing with these methods, our TaxoClass framework does not require labeled data for a set of seen classes.

Hierarchical Text Classification. Hierarchical text classification leverages a class hierarchy to improve the standard text classification performance. Typical methods can be divided into two categories: (1) *local approaches* which learn a text classifier per class [1], per parent class [17], or per level [30], and (2) *global approaches* which incorporate taxonomy structure information into one single classifier through recursive regularization [5] or graph neural network (GNN) based encoder [8, 24, 36]. Our TaxoClass framework adopts the global approach and uses a GNN-based encoder to obtain each class's representation.

5.3 TaxoClass: Weakly-Supervised Hierarchical Multi-label Text Classification

In this section, we first introduce our notations and task definition. Then, we present our TaxoClass framework which consists of four major steps: (1) document-class similarity calculation, (2) document core class mining, (3) core class guided classifier training, and (4) multi-label self-training. Figure 5.2 shows our framework overview and below sections discuss each step in more details.

5.3.1 Problem Formulation

Notations. A *corpus* $\mathcal{D} = \{D_1, \ldots, D_N\}$ is a text collection where each document $D_i \in \mathcal{D}$ is a sequence of words. A *class taxonomy* $\mathcal{T} = (\mathcal{C}, \mathcal{R})$ is a directed acyclic graph where each node represents a class c_j and each directed edge $\langle c_m, c_n \rangle \in \mathcal{R}$ indicates that parent class c_m is more general than the child class c_n. In this work, we assume each class c_j has a surface name s_j (either a word or a phrase) that serves as the weak supervision signal.

Task Definition. Given an unlabeled corpus $\mathcal{D}$, a class hierarchy $\mathcal{T} = (\mathcal{C}, \mathcal{R})$, and class surface names $\mathcal{S} = \{s_j\}_{j=1}^{|\mathcal{C}|}$, our task is to learn a text classifier $f(\cdot)$ that maps a new document D_{new} to its target $\mathbf{y} = [y_1, \ldots, y_{|\mathcal{C}|}] \in \mathcal{Y} = \{0, 1\}^{|\mathcal{C}|}$ where y_j equals to 1 if this document is categorized with class c_j and 0 otherwise.

Discussion. When the number of classes $|\mathcal{C}|$ is large (as it is in many HMTC applications), we can no longer assume all class surface names in $\mathcal{S}$ will explicitly appear in the given corpus $\mathcal{D}$ as done in most previous studies [15, 21, 29]. This is because many class names are actually summarizing phrases provided by humans (e.g., "*grocery and gourmet food*" in Fig. 5.1). As a result, we need to design a method that works under such a scenario.

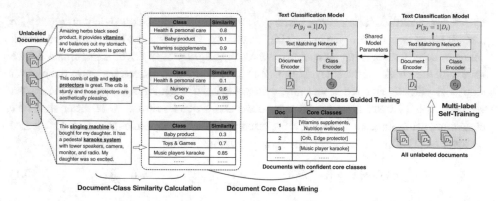

Fig. 5.2 Our TaxoClass framework overview. We first calculate document-class similarities using a textual entailment model (Sect. 5.3.2). Then, we identify document core classes (Sect. 5.3.3) and train a taxonomy-enhanced text classifier (Sect. 5.3.4). Finally, we generalize the classifier via multi-label self-training (Sect. 5.3.5). The "shared model parameters" indicates that we do self-training on the same model learned using our identified core classes

5.3.2 Document-Class Similarity Calculation

We take a textual entailment approach [33] to calculate the semantic similarity between each ⟨document, class⟩ pair. This approach imitates how humans determine whether a document is similar to a class or not—we read this document, create a hypothesis by filling the class name into a template (e.g., *"this document is about __"*), and ask ourselves to what extent this hypothesis is correct, given the context document.

In this work, we adopt a pre-trained textual entailment model that inputs a document D_i as the "premise", a template filled with a class name s_j as the "hypothesis", and outputs a probability of how likely this premise can entail the hypothesis. We treat this probability $\mathbf{P}(D_i \rightarrow c_j)$ as the document-class similarity $sim(D_i, c_j)$. More specifically, we use `Roberta-Large-MNLI`[1] as our textual entailment model which utilizes the pre-trained `Roberta-Large` as its backbone and is fine-tuned on the MNLI dataset.

5.3.3 Document Core Class Mining

When asked to tag a document with a set of classes from a class taxonomy, humans will first pinpoint a few classes that are most essential to this document. We refer to those most essential classes as the "core classes" and identify them in below two steps.

---

[1] https://huggingface.co/roberta-large-mnli.

Core Class Candidate Selection. We observe that on average each document is tagged with a small set of classes from the entire class taxonomy. Therefore, we first reduce the search space of core classes using a top-down approach (c.f. Fig. 5.3). Given a document D, we start with the "*Root*" class at level $l = 0$, find its two children classes that have the highest similarity with D, and add them into a queue. Then, for each class at level l in the queue, we select $l + 2$ classes from its children classes that are most similar to D. After all level l classes are processed, we aggregate all selected children classes and choose $(l + 1)^2$ classes (at level $l + 1$) with the highest path score (ps) defined below:

$$ps(Root) = 1, \quad ps(c_j) = \max_{c_k \in Par(c_j)} \{ps(c_k) \cdot sim(c_j, D)\}, \tag{5.1}$$

where $Par(c_j)$ is class c_j's parent class set. All chosen classes (at level $l + 1$) will be pushed into the queue and we stop this process when no class in the queue has further children. Finally, all classes that have entered the queue, except for the "*Root*" class, consist of the core class candidate set, denoted as $\mathbb{C}_i^{cand}$ for document D_i.

Confident Core Class Identification. For each document, we identify its core classes from the above selected candidate set based on two observations. First, a document usually has higher similarity with its core class c than with the parent and sibling classes of c. Take the document D_2 in Fig. 5.2 as an example, the similarity between D_2 and its core class "*crib*" is 0.95, much higher than the similarity between D_2 and core class's parent class "*nursery*" (0.6) as well as core class's sibling classes. Based on this observation, we define the "confidence score" of a candidate core class c for a document D as below:

$$conf(D, c) = sim(D, c) - \max_{c' \in Par(c) \cup Sib(c)} \{sim(D, c')\}, \tag{5.2}$$

where $Sib(c)$ represents the sibling class set of c.

Our second observation is that the similarity between a document D and its core class c is salient from a *corpus-wise* perspective. Namely, if a class c is a document D's core class, the confidence score $conf(D, c)$ is higher than the median confidence score[2] between class c and all documents tagged with c (denoted as $\mathcal{D}(c)$). Formally, we have:

$$conf(D, c) \geq median\{conf(D', c) | D' \in \mathcal{D}(c)\}. \tag{5.3}$$

According to this observation, we check each class in document D_i's candidate core set $\mathbb{C}_i^{cand}$ and add classes that satisfy the above criteria into the final core class set $\mathbb{C}_i$. Note here this core class set $\mathbb{C}_i$ could be empty when document D_i does not have any confident core class.

---

[2] We have also tried using "average" but empirically found that using "median" is better and more robust to outliers.

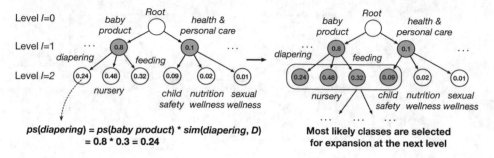

ps(diapering) = ps(baby product) * sim(diapering, D)
         = 0.8 * 0.3 = 0.24

Most likely classes are selected
for expansion at the next level

Fig. 5.3 Top-down core class candidate selection

5.3.4 Core Class Guided Classifier Training

Based on identified document core classes, we train one classifier for hierarchical multi-label text classification. Below we first introduce our classifier architecture and then present our training method.

Text Classifier Architecture. We design our classifier to have a dual-encoder architecture: one document encoder maps document D_i to its representation $\mathbf{D}_i$, one class encoder learns class c_j's representation $\mathbf{c}_j$, and one matching network returns the probability of document D_i being tagged with class c_j.

We instantiate our document encoder $g_{\text{doc}}(\cdot)$ to be a pre-trained BERT-base-uncased model [4] and follow previous work [22] to use the [CLS] token representation as the document representation. For class encoder $g_{\text{class}}(\cdot)$, we follow [26] and use a graph neural network (GNN) [12] to model the class taxonomy structure. This taxonomy-enhanced class encoder can capture both the textual information from class surface names and structural information from the class taxonomy.

Given a class c_j, we first obtain its ego network that includes its parent and children classes in the class taxonomy, as shown in Fig. 5.4. Then, we input this ego network to a GNN that propagates node features over the network structure. The node features are initialized with the pre-trained word embeddings of class surface names.[3] The propagation mechanism updates the feature of a node u by iteratively aggregating representations of its neighbors and itself. Formally, we define a GNN with L-layers as follows:

$$h_u^{(l)} = \text{ReLU}\left(\sum_{v \in N(u)} \alpha_{uv}^{(l-1)} \mathbf{W}^{(l-1)} h_v^{(l-1)} \right), \qquad (5.4)$$

where $l \in \{1, \ldots, L\}$, $N(u)$ includes node u's neighbors and itself, $\alpha_{uv}^{(l-1)} = \frac{1}{\sqrt{|N(u)||N(v)|}}$ is a normalization constant (same for all layers), and $\mathbf{W}^{(l-1)}$ are learnable parameters.

---

[3] For multi-gram class names, we use their averaged word embeddings.

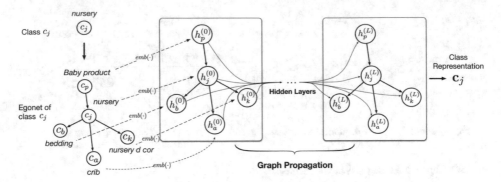

Fig. 5.4 Taxonomy-enhanced class encoder in TaxoClass

After obtaining individual node features, we combine them into a vector representing the whole ego network G as follows:

$$h_G = \frac{1}{|G|} \sum_{u \in G} h_u^{(L)}. \tag{5.5}$$

As this ego network is centered on class c_j and encodes its both textual and structural information, we treat this final graph representation as the class representation $\mathbf{c}_j$.

Based on the document representation $\mathbf{D}_i$ and the class representation $\mathbf{c}_j$, we use a log-bilinear text matching model to compute the probability of document D_i being tagged with class c_j as follows:

$$p_{ij} = \mathbf{P}(y_j = 1 | D_i) = \sigma(\exp(\mathbf{c}_j^T \mathbf{B} \mathbf{D}_i)), \tag{5.6}$$

where $\sigma(\cdot)$ is the sigmoid function and $\mathbf{B}$ is a learnable interaction matrix.

Text Classifier Training. We use our discovered document core classes to train a text classifier. One strategy is to treat each document's core classes as positive classes and all the remaining classes as negative classes. However, this strategy has a high false negative rate because some non-core classes could still be relevant to the document (c.f. Fig. 5.1).

We observe a document's multiple labeled classes usually have some ancestor-descendent relations in the class hierarchy $\mathcal{T} = (\mathcal{C}, \mathcal{R})$. This implies that given a document's core class, its parent class and some of its children classes are also likely to be tagged with this document. Therefore, we introduce all core classes' parent classes into the positive class set and exclude their children classes from the negative class set. Formally, given a document D_i with its core class set $\mathbb{C}_i$, we define its positive and negative class set as follows:

$$\mathbb{C}_i^{pos} = \left(\bigcup_{c_j \in \mathbb{C}_i} Par(c_j) \right) \cup \mathbb{C}_i, \quad \mathbb{C}_i^{neg} = \mathcal{C} - \mathbb{C}_i^{pos} - \bigcup_{c_j \in \mathbb{C}_i} Chd(c_j), \tag{5.7}$$

where $Chd(c_j)$ is class c_j's children class set. Finally, we train our classification model using the below binary cross entropy (BCE) loss:

$$\mathcal{L} = -\sum_{\substack{i=1 \\ \mathbb{C}_i \neq \emptyset}}^{|\mathcal{D}|} (\sum_{c_j \in \mathbb{C}_i^{pos}} \log p_{ij} + \sum_{c_j \in \mathbb{C}_i^{neg}} \log(1 - p_{ij})), \qquad (5.8)$$

where "Ø" indicates an empty set and we exclude the documents without any confident core class from the loss calculation.

5.3.5 Multi-label Self-training

After training the text classifier based on document core classes, we propose to further refine the model via self-training on the entire unlabeled corpus $\mathcal{D}$ for better generalization. The idea of self-training (ST) [32] is to iteratively use the model's current prediction P to compute a target distribution Q which guides the model for refinement. In general, the ST objective is expressed with the KL divergence loss as below:

$$\mathcal{L}_{ST} = \text{KL}(Q||P) = \sum_{i=1}^{|\mathcal{D}|} \sum_{j=1}^{|\mathcal{C}|} q_{ij} \log \frac{q_{ij}}{p_{ij}}. \qquad (5.9)$$

The target distribution Q is constructed by enhancing high-confidence predictions while down-weighting low-confidence ones:

$$q_{ij} = \frac{p_{ij}^2/(\sum_i p_{ij})}{p_{ij}^2/(\sum_i p_{ij}) + (1 - p_{ij})^2/(\sum_i (1 - p_{ij}))}. \qquad (5.10)$$

Different from the previous studies [20], our target distribution Q can be applied to multi-label classification problem as it normalizes the current predictions P for each individual class. Intuitively, this equation can enhance high-confidence predictions while down-weighting low-confidence predictions. This is because if example i is more confidently labeled with class j than other examples, we will have a large p_{ij} that dominates the $\sum_i p_{ij}$ term. Consequently, Eq. (5.10) computes a large q_{ij}, which pushes the model to predict class j for example i.

In practice, instead of updating the target distribution Q for every training example, we update it every 25 batches[4] and train the model with Eq. (5.9), which makes the self-training process more efficient and robust. We summarize our TaxoClass framework in Algorithm 5.1.

---

[4] This hyper-parameter controls the update frequency. Empirically, we find our model is insensitive to this hyper-parameter (in the typical value range of 10–100).

Algorithm 5.1: TaxoClass Framework

   Input: An unlabeled corpus $\mathcal{D}$, a class taxonomy $\mathcal{T}$ with class names $\mathcal{S}$, an entailment model
        $\mathcal{M}$, total number of batches B.
   Output: A trained classifier $f(\cdot)$.
1 Use model $\mathcal{M}$ to compute document-class similarity (c.f. Sect. 5.3.2);
2 Obtain document core classes $\{(D_i, \mathbb{C}_i) \mid D_i \in \mathcal{D}\}$ (c.f. Sect. 5.3.2);
3 Train classifier $f(\cdot)$ with Eq. (5.8);
4 **for** i *from 1 to* B **do**
5   **if** i *mod 25 = 0* **then**
6   Update Q with Eq. (5.10);
7   Train classifier $f(\cdot)$ with Eq. (5.9);
8 Return $f(\cdot)$;

5.4   Experiments

5.4.1 Datasets

We use two public datasets from different domains to evaluate our method: (1) **Amazon-531** [18] contains 49,145 product reviews and a three-level class taxonomy consisting of 531 classes; and (2) **DBPedia-298** [14] includes 245,832 Wikipedia articles and a three-level class taxonomy with 298 classes. Documents in both datasets are lower-cased and truncated to has maximum 500 tokens. We list the data statistics in Table 5.1.

5.4.2 Compared Methods

To the best of our knowledge, we are the first to study *weakly-supervised* HMTC problem and there is no directly comparable baseline under the exact same setting as ours. Therefore, we choose a wide range of representative methods that are most related to TaxoClass and adapt them to our problem setting, described as follows.

Table 5.1 Dataset statistics. Supervised methods are trained on the entire training set. Weakly-supervised methods are trained by treating the training set as unlabeled data. All methods are evaluated on the test set

Dataset	# Train	# Test	# Classes
Amazon-531	29,487	19,685	531
DBPedia-298	196,665	49,167	298

- **Hier-doc2vec** [13][5]: This *weakly-supervised* method first embeds documents and classes into a shared semantic space, and then recursively selects the class of the highest embedding similarity with the document in a top-down fashion. We set the embedding dimensionality to be 100 and use the default value for all other hyper-parameters.[6]

- **WeSHClass** [21][7]: Another *weakly-supervised* method that generates pseudo documents to pre-train a text classifier and bootstraps the pre-trained classifier on unlabeled documents with self-training. The class surface names are treated as the "class-related keywords" in this method. For the pseudo document generation step, we use its internal LSTM language model. We treat all classes in its returned class path as the output classes.

- **SS-PCEM** [31][8]: This *semi-supervised* method uses a generative model to generate documents based on a class path sampled from the class taxonomy. Both labeled and unlabeled documents are used to fit this generative model via the EM algorithm. Finally, it uses the posterior probability of a test document to predict its labeled classes. Among different base classifiers, we choose their author reported best variant PCEM in this study. We use 30% of labeled training documents for this method.

- **Hier-0Shot-TC** [33][9]: This *zero-shot* method uses a pre-trained textual entailment model to predict to what extent a document (as the premise text) can entail a template filled with the class name (as the hypothesis text). Similar to **Hier-doc2vec**, we select the class with the highest entailment score at each level in a top-down recursive fashion. For fair comparison, we change its internal `BERT-base-uncased` model to `RoBERTa-large-mnli` model as is used in our method.

- **TaxoClass**[10]: Our proposed *weakly-supervised* framework that identifies document core classes, leverages core classes to train a taxonomy-enhanced text classifier, and generalizes the classifier using multi-label self-training. We also evaluate two ablations: **TaxoClass-NoST** which removes the multi-label self-training step, and **TaxoClass-NoGNN** which replaces the GNN-based class encoder with a simple embedding layer initialized with pre-trained word embeddings.

---

[5] https://radimrehurek.com/gensim/models/doc2vec.html.

[6] We also test the Flat-doc2vec variant which directly ranks all classes in the taxonomy and returns top ranked classes. Its performance is significantly worse than Hier-doc2vec and thus we only report Hier-doc2vec results.

[7] https://github.com/yumeng5/WeSHClass.

[8] https://github.com/HKUST-KnowComp/PathPredictionForTextClassification.

[9] https://github.com/yinwenpeng/BenchmarkingZeroShot.

[10] https://github.com/mickeystroller/TaxoClass.

5.4.3 Evaluation Metrics

We follow previous studies [9, 34] and evaluate the multi-label classification results from different aspects using various metrics. The first metric is **Example-F1**[11] which calculates the average F1 scores for all documents as follows:

$$\text{Example-F1} = \frac{1}{N} \sum_{i=1}^{N} \frac{2|\mathbb{C}_i^{true} \cap \mathbb{C}_i^{pred}|}{|\mathbb{C}_i^{true}| + |\mathbb{C}_i^{pred}|}, \tag{5.11}$$

where $\mathbb{C}_i^{true}$ ($\mathbb{C}_i^{pred}$) is the true (model predicted) class set of document D_i.

Moreover, as many applications formalize the HMTC as a class ranking problem [6, 9], we convert predicted class set $\mathbb{C}_i^{pred}$ into a rank list $\mathbb{R}_i^{pred}$ based on each class's model predicted probability and calculate **Precision at k ($P@k$)** as follows:

$$P@k = \frac{1}{N} \sum_{i=1}^{N} \frac{|\mathbb{C}_i^{true} \cap \mathbb{R}_{i,1:k}^{pred}|}{min(k, |\mathbb{C}_i^{true}|)}, \tag{5.12}$$

where $\mathbb{R}_{i,1:k}^{pred}$ is each method predicted top k most likely classes for D_i. Finally, for methods able to return the probability of a document being tagged with each class in the taxonomy, we calculate their **Mean Reciprocal Rank (MRR)** as follows:

$$\text{MRR} = \frac{1}{N} \sum_{i=1}^{N} \frac{1}{|\mathbb{C}_i^{true}|} \sum_{c_j \in \mathbb{C}_i^{true}} \frac{1}{R_{ij}}, \tag{5.13}$$

where R_{ij} is the "rank" of document D_j's true class c_j in model predicted rank list.

5.4.4 Implementation Details

For all baseline methods except Hier-doc2vec, we use the public implementations from their authors and leave the hyper-parameters unchanged. For both Hier-0Shot-TC and our method, we adopt the same public `Roberta-Large-MNLI` model as the textual entailment model and use the same hypothesis template: "*this product is about __ .*" for Amazon-531 dataset and "*this example is __.*" for DBPedia-298 dataset. We use AdamW optimizer to train our model with batch size 64, learning rate 5e-5 for all parameters in BERT document encoder and learning rate 4e-3 for all remaining parameters. During the multi-label self-training stage, we use learning rate 1e-6 for all parameters in the BERT document encoder and 5e-4 for all remaining parameters. We run all experiments on a single cluster with 80 CPU cores and a Quadro RTX 8000 GPU. All deep learning models are moved to the GPU for faster inference speed. With batch size 64, the TaxoClass framework consumes about 10GB GPU memory. In principle, all methods should be runnable on CPU.

---

[11] This metric is also called "micro-Dice coefficient".

Table 5.2 Evaluation of all compared text classification methods on two datasets. For some methods predicting a class path in a top-down fashion rather than returning all classes' probabilities, we cannot compute their MRR scores and indicate this using "N/A"

Method	Amazon-531				DBPedia-298			
	Example-F1	P@1	P@3	MRR	Example-F1	P@1	P@3	MRR
Hier-doc2vec [13]	0.3157	0.5805	0.3115	N/A	0.1443	0.2635	0.1443	N/A
WeSHClass [21]	0.2458	0.5773	0.2517	N/A	0.3047	0.5359	0.3048	N/A
TaxoClass-NoST	0.5431	0.7918	0.5414	0.5911	0.7712	0.8621	0.7712	0.8221
TaxoClass-NoGNN	0.5271	0.7642	0.5213	0.5621	0.7241	0.8154	0.7241	0.7692
TaxoClass	**0.5934**	**0.8120**	**0.5894**	**0.6332**	**0.8156**	**0.8942**	**0.8156**	**0.8762**
SS-PCEM [31]	0.2921	0.5369	0.2948	0.3004	0.3845	0.7424	0.3845	0.4032
Hier-0Shot-TC [33]	0.4742	0.7144	0.4610	N/A	0.6765	0.7871	0.6765	N/A

5.4.5 Overall Performance Comparison

Table 5.2 presents the overall results of all compared methods. First, we find most weakly-supervised (i.e., WeSHClass, TaxoClass and its variants) and zero-shot method (i.e., Hier-0Shot-TC) can outperform the semi-supervised method SS-PCEM even the later has access to 30% of labeled documents. Second, we can see that TaxoClass has the overall best performance across all the metrics and defeats the second best method by a large margin. Comparing TaxoClass with TaxoClass-NoGNN, we show the importance of incorporating taxonomy structure into the class encoder. Moreover, the improvement of TaxoClass over TaxoClass-NoST demonstrates the effectiveness of our multi-label self-training.

5.4.6 Effectiveness of Core Class Mining

We evaluate the effectiveness of our core class mining method as follows. First, we define a set of rival methods and use them to generate various sets of "core classes". Then, we derive pseudo-training data for each generated core class set and use it to learn a text classifier with the same architecture as the one in TaxoClass. Finally, we report each model's performance on the test set. Note here we skip the self-training step to ensure the "core class based pseudo-training data" is the only variable.

Table 5.3 Evaluation of core class mining algorithms on Amazon-531 dataset. We train the classifier using different training sets derived from different core class mining algorithm outputs

Core class mining method	Example-F1	P@1	P@3	MRR
Explicit Mention	0.1611	0.2168	0.1564	0.2045
0Shot	0.4793	0.7361	0.4782	N/A
Ours	**0.5431**	**0.7918**	**0.5414**	**0.5911**
Ours-NoCS	0.3812	0.6254	0.3831	0.4366
Ours-NoConf	0.2603	0.4431	0.2521	0.3014

Table 5.3 lists all the results. First, we find that the "Explicit Mention" method, which treats all classes with names explicitly appear in the corpus as the core classes, does not perform well for our HMTC problem. One reason could be many class names are human-curated summarizing phrases that do not appear in the corpus naturally. Second, the "0Shot" method views the output classes of baseline method Hier-0Shot-TC as the core classes and trains a new classifier. Interestingly, this new classifier performs better than the original Hier-0Shot-TC classifier, which shows that transferring knowledge from a general zero-shot classifier to a domain-specific classifier is a possible and promising direction. Finally, we compare variants of our own methods. The "Ours-NoCS" method removes the candidate core class selection step and treats all classes with high confidence scores as core classes. The "Ours-NoConf" method skips the confident core class identification step and views all candidate core classes as the final output core classes. We can see a significant performance drop on both ablations, which shows the importance of our two core class mining steps.

5.4.7 Analysis of Classifier Architecture

We study whether we can use the identified document core classes to train other text classifiers with different architectures such as fastText [10] and TextCNN [11]. As shown in Table 5.4, both methods achieve reasonable performance. We can also see that TaxoClass with and without GNN-enhanced class encoder can outperform both methods. This shows the effectiveness of our dual-encoder classifier architecture.

5.4.8 Supervision Signals in Class Names

We vary the percentage of labeled documents on Amazon-531 dataset for training a supervised fastText classifier and present its corresponding performance in Fig. 5.5. We can see the performance of our TaxoClass framework is equivalent to that of a supervised fastText model learned using roughly 70% of labeled documents in the training set (i.e., about 20,000 labeled documents).

Table 5.4 Performance of different classifiers on Amazon-531 dataset. All methods use the same training set derived from our identified document core classes

Method	Example-F1	P@1	P@3	MRR
fastText	0.4472	0.7515	0.4521	0.4587
TextCNN	0.4787	0.7694	0.4771	0.4827
TaxoClass-NoGNN	0.5271	0.7642	0.5213	0.5621
TaxoClass	0.5934	0.8120	0.5894	0.6332

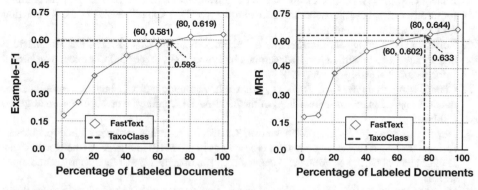

Fig. 5.5 Comparison between TaxoClass and supervised fastText method on Amazon-531 dataset. We train the fastText model using on different percentages of labeled training documents

5.5 Summary

In this chapter, we study the hierarchical multi-label text classification problem when only class surface names, instead of massive labeled documents, are given. We propose a novel TaxoClass framework which leverages the class taxonomy to derive document core classes and learns taxonomy-enhanced text classifier for prediction. Extensive experiments demonstrate the effectiveness of TaxoClass on two real-world datasets from different domains.

In the future, we plan to explore how TaxoClass framework can be integrated with semi-supervised methods and data augmentation methods, when some class surface names are too ambiguous to indicate class semantics. Another another line, we may also interpret the document structuring task as a weakly-supervised hierarchical clustering problem where taxonomy nodes are viewed as seed guidances and potential cluster centers. Then, we can simultaneously cluster both documents and keywords to construct a document-allocated topic taxonomy. Finally, we consider extending our multi-label self-training method to other related NLP tasks such as fine-grained entity typing.

References

1. Banerjee, S., Akkaya, C., Perez-Sorrosal, F., Tsioutsiouliklis, K.: Hierarchical transfer learning for multi-label text classification. In: Proceedings of the 57th Annual Meeting of the Association for Computational Linguistics (2019)
2. Berthelot, D., Carlini, N., Goodfellow, I., Papernot, N., Oliver, A., Raffel, C.: Mixmatch: a holistic approach to semi-supervised learning. In: Proceedings of the 33rd Conference on Neural Information Processing Systems (2019)
3. Chang, M.W., Ratinov, L.A., Roth, D., Srikumar, V.: Importance of semantic representation: dataless classification. In: Proceedings of the 2008 AAAI Conference on Artificial Intelligence (2008)
4. Devlin, J., Chang, M.W., Lee, K., Toutanova, K.: BERT: Pre-training of deep bidirectional transformers for language understanding. In: Proceedings of the 2019 Conference of the North American Chapter of the Association for Computational Linguistics: Human Language Technologies (2019)
5. Gopal, S., Yang, Y.: Recursive regularization for large-scale classification with hierarchical and graphical dependencies. In: Proceedings of the 19th ACM SIGKDD International Conference on Knowledge Discovery and Data Mining (2013)
6. Guo, C.F., Mousavi, A., Wu, X., Holtmann-Rice, D.N., Kale, S., Reddi, S.J., Kumar, S.: Breaking the glass ceiling for embedding-based classifiers for large output spaces. In: Proceedings of the 32nd Conference on Neural Information Processing Systems (2019)
7. Gururangan, S., Dang, T., Card, D., Smith, N.A.: Variational pretraining for semi-supervised text classification. In: Proceedings of the 57th Annual Meeting of the Association for Computational Linguistics (2019)
8. Huang, W., Chen, E., Liu, Q., Chen, Y., Huang, Z., Liu, Y., Zhao, Z., Zhang, D., Wang, S.: Hierarchical multi-label text classification: An attention-based recurrent network approach. In: Proceedings of the 28th ACM International Conference on Information and Knowledge Management (2019)
9. Jain, H., Prabhu, Y., Varma, M.: Extreme multi-label loss functions for recommendation, tagging, ranking, and other missing label applications. In: Proceedings of the 22nd ACM SIGKDD International Conference on Knowledge Discovery and Data Mining (2016)
10. Joulin, A., Grave, E., Bojanowski, P., Mikolov, T.: Bag of tricks for efficient text classification. In: Proceedings of the 15th Conference of the European Chapter of the Association for Computational Linguistics (2016)
11. Kim, Y.: Convolutional neural networks for sentence classification. In: Proceedings of the 2014 Conference on Empirical Methods in Natural Language Processing (2014)
12. Kipf, T.N., Welling, M.: Semi-supervised classification with graph convolutional networks. In: Proceedings of the 5th International Conference on Learning Representations (2017)
13. Le, Q.V., Mikolov, T.: Distributed representations of sentences and documents. In: Proceedings of the 31st International Conference on Machine Learning (2014)

14. Lehmann, J., Isele, R., Jakob, M., Jentzsch, A., Kontokostas, D., Mendes, P.N., Hellmann, S., Morsey, M., Van Kleef, P., Auer, S., et al.: DBpedia—a large-scale, multilingual knowledge base extracted from Wikipedia. Semant. Web **6**(2), 167–195 (2015)
15. Li, K., Li, S., Yavuz, S., Zha, H., Su, Y., Yan, X.: Hiercon: hierarchical organization of technical documents based on concepts. In: Proceedings of the 19th IEEE International Conference on Data Mining (2019)
16. Li, K., Zha, H., Su, Y., Yan, X.: Unsupervised neural categorization for scientific publications. In: Proceedings of the 2018 SIAM International Conference on Data Mining (2018)
17. Liu, T., Yang, Y., Wan, H., Zeng, H., Chen, Z., Ma, W.: Support vector machines classification with a very large-scale taxonomy. In: Proceedings of the 11st ACM SIGKDD International Conference on Knowledge Discovery and Data Mining (2005)
18. McAuley, J.J., Leskovec, J.: Hidden factors and hidden topics: understanding rating dimensions with review text. In: Proceedings of the 7th ACM conference on Recommender Systems (2013)
19. Mekala, D., Shang, J.: Contextualized weak supervision for text classification. In: Proceedings of the 58th Annual Meeting of the Association for Computational Linguistics (2020)
20. Meng, Y., Shen, J., Zhang, C., Han, J.: Weakly-supervised neural text classification. In: Proceedings of the 27th ACM International Conference on Information and Knowledge Management (2018)
21. Meng, Y., Shen, J., Zhang, C., Han, J.: Weakly-supervised hierarchical text classification. In: Proceedings of the 2019 AAAI Conference on Artificial Intelligence (2019)
22. Meng, Y., Zhang, Y., Huang, J., Xiong, C., Ji, H., Zhang, C., Han, J.: Text classification using label names only: A language model self-training approach. In: Proceedings of the 2020 Conference on Empirical Methods in Natural Language Processing (2020)
23. Nam, J., Mencía, E.L., Fürnkranz, J.: All-in text: learning document, label, and word representations jointly. In: Proceedings of the 2016 AAAI Conference on Artificial Intelligence (2016)
24. Peng, H., Li, J., He, Y., Liu, Y., Bao, M., Wang, L., Song, Y., Yang, Q.: Large-scale hierarchical text classification with recursively regularized deep graph-cnn. In: Proceedings of the 2018 World Wide Web Conference (2018)
25. Rios, A., Kavuluru, R.: Few-shot and zero-shot multi-label learning for structured label spaces. In: Proceedings of the 2018 Conference on Empirical Methods in Natural Language Processing (2018)
26. Shen, J., Shen, Z., Xiong, C., Wang, C., Wang, K., Han, J.: TaxoExpan: self-supervised taxonomy expansion with position-enhanced graph neural network. In: Proceedings of the 2020 Web Conference (2020)
27. Song, Y., Roth, D.: On dataless hierarchical text classification. In: Proceedings of the 2014 AAAI Conference on Artificial Intelligence (2014)
28. Srivastava, S., Labutov, I., Mitchell, T.M.: Zero-shot learning of classifiers from natural language quantification. In: Proceedings of the 56th Annual Meeting of the Association for Computational Linguistics (2018)
29. Wang, Z., Mekala, D., Shang, J.: X-class: text classification with extremely weak supervision. In: Proceedings of the 2021 Conference of the North American Chapter of the Association for Computational Linguistics: Human Language Technologies (2021)
30. Wehrmann, J., Cerri, R., Barros, R.C.: Hierarchical multi-label classification networks. In: Proceedings of the 35th International Conference on Machine Learning (2018)
31. Xiao, H., Liu, X., Song, Y.: Efficient path prediction for semi-supervised and weakly supervised hierarchical text classification. In: Proceedings of the 2019 Web Conference (2019)
32. Xie, J., Girshick, R.B., Farhadi, A.: Unsupervised deep embedding for clustering analysis. In: Proceedings of the 33rd International Conference on Machine Learning (2016)

33. Yin, W., Hay, J., Roth, D.: Benchmarking zero-shot text classification: datasets, evaluation and entailment approach. In: Proceedings of the 2019 Conference on Empirical Methods in Natural Language Processing (2019)

34. You, R., Dai, S., Zhang, Z., Mamitsuka, H., Zhu, S.: Attentionxml: extreme multi-label text classification with multi-label attention based recurrent neural networks. In: Proceedings of the 33rd Conference on Neural Information Processing Systems (2019)

35. Zeng, Z., Zhou, W., Liu, X., Song, Y.: A variational approach to weakly supervised document-level multi-aspect sentiment classification. In: Proceedings of the 2019 Conference of the North American Chapter of the Association for Computational Linguistics: Human Language Technologies (2019)

36. Zhou, J., Ma, C., Long, D., Xu, G., Ding, N., Zhang, H., Xie, P., Liu, G.: Hierarchy-aware global model for hierarchical text classification. In: Proceedings of the 58th Annual Meeting of the Association for Computational Linguistics (2020)

Conclusions

6

6.1 Summary

In this book, we have proposed a principled framework that automatically constructs, enriches, and applies taxonomies for unleashing hidden knowledge in unstructured text. The whole automated framework requires minimum human labeled data and obtains good performances by (1) leveraging user-provided seed information as weak supervision, (2) utilizing salient statistical signals in unlabeled data as self supervision, and (3) resorting to existing knowledge repositories as distant supervision. Those obtained taxonomies, either a concept taxonomy or an event taxonomy, not only contain useful knowledge in themselves but also can be used to organize and index knowledge, help users search and comprehend knowledge stored in a large corpus, and provide structured guidance in many other knowledge engineering tasks.

6.2 Future Work

With the mounting big data and diverse applications in today's information-based society, my proposed taxonomy-centric framework will play an increasingly important role in organizing concepts, data, and knowledge. My long-term research goal is to create data-driven methods that ingest massive heterogeneous data, organize machine-actionable knowledge into taxonomies, and utilize taxonomies to facilitate human decision-making. Many unique challenges arise in this context and call for collaborative research efforts from multiple areas, including data mining, machine learning, natural language processing, computer vision, human-computer interaction, computer security, and much more. Below are some specific directions that I am excited to explore in the near future.

© The Author(s), under exclusive license to Springer Nature Switzerland AG 2022
J. Shen and J. Han, *Automated Taxonomy Discovery and Exploration*,
Synthesis Lectures on Data Mining and Knowledge Discovery,
https://doi.org/10.1007/978-3-031-11405-2_6

6.2.1   Integrate Heterogeneous Modalities and Sources

While my current research primarily focuses on text data, recent years have witnessed a trend in the confluence of multiple data modalities (e.g., graphs, image, time series) gathered from heterogeneous sources. On one hand, introducing multi-modal data opens the gate to more labeled datasets for training models. On the other hand, human's understanding is often grounded on multiple data modalities. A multi-modal multi-task system can reflect a better understanding of the world. Some study [1] has already shown that more meaningful taxonomic relations can be recovered from text-rich networks (e.g., a citation network with paper abstracts, a social network with user posts, etc.) from their corresponding corpora without metadata. Furthermore, people find that leveraging the information in taxonomy can help learn better representations of knowledge graph (modeled as a heterogeneous information network) [2]. My proposed framework, if properly extended, could well accommodate such heterogeneity. I plan to develop novel embedding methods that organize heterogeneous data into a shared latent space for bridging different modalities.

6.2.2   Engage with Human Behaviors and Interactions

For many complex analytical tasks, humans and machines need to collaborate to acquire necessary task-specific knowledge. There is great potential to adopt my proposed taxonomy-centric framework to facilitate such a human-in-the-loop process: (1) machines input user-provided seed data as weak supervision and task guidance, perform data analysis by learning task-specific models, and return interpretable patterns and visualizations; and (2) humans make sense of the resultant patterns and visual cues, adjust or provide additional seed data, and give feedback to guide the machines to extract more useful knowledge. I am interested in working with researchers in HCI, visualization, and machine learning to address fundamental challenges for realizing this goal. Example research problems include: How to design intuitive interfaces to help users provide seed data for various applications? How to train a predictive and interpretable model using limited labeled data from the user input? How to develop taxonomy-tailored visualization techniques to help users more easily gain useful knowledge? How to update a machine learning model continuously based on user feedback to better satisfy users' information needs? To answer some of the above questions, I plan to leverage the prompt-based few-shot learning techniques to incorporate sparse example-based user feedbacks into our current models.

6.2.3 Preserve Data Privacy and Model Security

While the knowledge discovery process from multiple data sources through collective learning is powerful, the uncensored information extraction and exchanging of multi-party data may be prone to adversarial attacks and threaten users' privacy. For example, an online retailer may provide false or exaggerated product descriptions to fool a category-guided recommender system to rank their products in top positions. An identity thief may conduct differential privacy attacks to infer sensitive user information, which is extremely dangerous and concerning. My future work on taxonomy will also study privacy-preserving data mining methods and federated learning techniques to prevent such attacks. Several research directions include: (1) enforcing the differential privacy principle to select and anonymize personal identifiable features for achieving the optimal trade-off between model utility and data privacy, and (2) leverage the multi-party computation techniques to achieve secure data and knowledge exchange and prevent information leakage propagation.

References

1. Shi, Y., Shen, J., Li, Y., Zhang, N., He, X., Lou, Z., Zhu, Q., Walker, M., Kim, M.H., Han, J.: Discovering hypernymy in text-rich heterogeneous information network by exploiting context granularity. In: Proceedings of the 28th ACM International Conference on Information and Knowledge Management (2019)
2. Xiao, H., Song, Y.: Manifold alignment across geometric spaces for knowledge base representation learning. In: Proceedings of the 3rd Conference on Automated Knowledge Base Construction (2021)